BATTLING CLIMATE CHANGE AND TRANSFORMING AGRI-FOOD SYSTEMS

ASIA–PACIFIC RURAL DEVELOPMENT AND FOOD SECURITY FORUM 2022 HIGHLIGHTS AND TAKEAWAYS

DECEMBER 2022

ASIAN DEVELOPMENT BANK

ADB

© 2022 Asian Development Bank
6 ADB Avenue, Mandaluyong City, 1550 Metro Manila, Philippines
Tel +63 2 8632 4444; Fax +63 2 8636 2444
www.adb.org

Some rights reserved. Published in 2022.

ISBN 978-92-9270-002-7 (print); 978-92-9270-003-4 (electronic); 978-92-9270-004-1 (ebook)
Publication Stock No. SGP220608-2
DOI: http://dx.doi.org/10.22617/SGP220608-2

The views expressed in this publication are those of the authors and do not necessarily reflect the views and policies of the Asian Development Bank (ADB) or its Board of Governors or the governments they represent.

ADB does not guarantee the accuracy of the data included in this publication and accepts no responsibility for any consequence of their use. The mention of specific companies or products of manufacturers does not imply that they are endorsed or recommended by ADB in preference to others of a similar nature that are not mentioned.

By making any designation of or reference to a particular territory or geographic area, or by using the term "country" in this publication, ADB does not intend to make any judgments as to the legal or other status of any territory or area.

Please contact pubsmarketing@adb.org if you have questions or comments with respect to content, or if you wish to obtain copyright permission for your intended use that does not fall within these terms, or for permission to use the ADB logo.

Corrigenda to ADB publications may be found at http://www.adb.org/publications/corrigenda.

Notes:
In this publication, "$" refers to United States dollars.
ADB recognizes "China" as the People's Republic of China, "Korea" as the Republic of Korea, and "Russia" as Russian Federation.

On the cover: (Clockwise from left) Elderly farmer, wearing a native hat and rain cape, weeding his rice field in Lucban, Quezon in the Philippines (photo by Al Benavente/ADB). ADB-financed Grant 0144: Sustainable Natural Resources and Productivity Enhancement Project and Loan 1949: Smallholder Development Project - Organic vegetable farmers in Boung Phao Village (photo by Ariel Javellana/ADB). Female vendors keep the shop at a wet market in Phuket town. Thailand's female population comprises 47% of the country's workforce, which makes up the highest percentage of working women in the Asia and Pacific region (photo by Lester Ledesma/ADB). ADB-financed Grant 0144: Sustainable Natural Resources and Productivity Enhancement Project and Loan 1949: Smallholder Development Project - Organic vegetable farmers in Boung Phao Village (photo by Ariel Javellana/ADB). Radhika Bika at her Purna poultry farm at Syangja district, Nepal (photo by Narenda Shrestha/ADB).

Contents

Table, Figures, and Box

Table

Figures

Box

About the RDFS Forum 2022

The Asia-Pacific Rural Development and Food Security Forum (RDFS Forum 2022) was conducted virtually on 22-24 March 2022. The theme of the forum was "Battling Climate Change through Sustainable Agri-food Systems," and convened about 1,000 innovators, experts, and thought leaders to

- Share perspective on the future of agri-food systems and culture amidst complex and evolving challenges;
- Explore new research, innovations, and technologies that can help build nature-positive food systems; and
- Forge partnerships and collaborations which will help mobilize finance for innovation, research, and business to promote food and nutrition security in the Asia and Pacific region.

The 3-day online event consisted of 11 sessions: a leaders' roundtable discussion, four technical sessions, three deep dive sessions, one special session, one knowledge product showcase session, and one session on actions and recommendations.

The four technical sessions focused on (i) digital technologies for agriculture; (ii) pathways to sustainable and inclusive food systems; (iii) intersector approaches to nutrition security; and (iv) financing green, resilient, and inclusive agriculture; and future crosscutting issues.

Meanwhile, the three deep-dive sessions discussed issues on integrated rural–urban development, urban farming, and alternative proteins to meet the growing global food demand.

This report summarizes the highlights and key takeaways of each session during the RDFS Forum 2022. The report provides a summary of the sessions and a reference for the proceedings during the forum.

#RDFS2022 is a flagship event of ADB's Rural Development and Food Security (Agriculture) Thematic Group, Sustainable Development and Climate Change Department. This is the third offering (bit.ly/3vRnJJ9) of the event following #RDFS2019 (bit.ly/3vMSX4p) and #FoodSecurityForum2016 (bit.ly/35BYqjV).

#RDFS2022 Event Page: bit.ly/3tL8HCp
#RDFS2022 Forum Note: bit.ly/3sCrlMT
#RDFS2022 Program: bit.ly/39ovLA2
#RDFS2022 Key Highlights: bit.ly/3HX3UCN

Abbreviations

ADB	Asian Development Bank
AIF	Africa Improved Foods
CH4	methane
COVID-19	coronavirus disease
CSA	climate-smart agriculture
DBM	double burden of malnutrition
DMC	developing member country
DSM	Royal Dutch State Mines
ENP	Pakistan's Ehsaas Nashonuma Project
FAO	Food and Agriculture Organization of the United Nations
GEP	gross ecosystem product
GHG	greenhouse gas
INCFF	Innovative Natural Capital Financing Facility
InVEST	Integrated Valuation of Ecosystem Services and Tradeoffs
MMS	multiple micronutrient supplement
MOAC	Ministry of Agriculture and Cooperatives (Thailand)
MOE	Ministry of Education (Thailand)
MOI	Ministry of Interior (Thailand)
MOPH	Ministry of Public Health (Thailand)
NCD	noncommunicable disease
NUPAP	Philippines' National Urban and Peri-Urban Agriculture Program
PRC	People's Republic of China
PxD	Precision Development
RDFS	Rural Development and Food Security
UNICEF	United Nations Children's Fund

Executive Summary

With the growing impact of climate change on agriculture and food, its threat to human existence can no longer be dismissed. This threat is global, but producers and consumers in less developed and emerging economies are more vulnerable due to their lack of resources and capacity to withstand it. Globally, aid agencies, government departments, think tanks, and development banks are alerting the world to this issue. Policy makers everywhere are united on one thing: they do not assume that food security is a given. There is an urgent need to assemble thoughts and actions.

This report reflects the program of the Asia-Pacific Rural Development and Food Security Forum (RDFS Forum 2022) and embraces four main themes: (i) sustainable and inclusive food systems and the contribution of digital technology; (ii) financing sustainable agriculture and natural capital; (iii) nutrition security and the double burden of malnutrition; and (iv) the rural–urban divide. The content that forum participants delivered is incorporated in this report. It conveys the ideas and experiences of the forum participants and identifies the key challenges and levers for change in a world that is threatened by climate change. The concluding section presents readers with suggested approaches to the challenges that the developing world faces.

The need to improve productivity in the agri-food (relating to the commercial production of food by farming) chain is an important message, but it is now an even bigger challenge because climate change is working against the productivity gains derived from science and technology that were previously a key part of agricultural economic development. Climate change significantly reduces the productivity of the agri-food system through its many impacts on weather, water availability, above average temperatures, etc. If climate smart policies are to be successful, agriculture and its associated processing and distribution activities must be transformed.

It is in the self-interest of producers, processors, and consumers to speed up adaptation to climate change and give priority to the transformation of the agri-food system. Food chains in Asia and the Pacific are highly vulnerable to climate change impacts. This is a win-win situation. Adaptive policies are technically feasible and are a necessary step to achieve sustainable food and nutrition security. However, governments, development banks, and their partners need to play their part so that the world makes the step from "theoretically feasible" to "field friendly." Private producers need to be involved—in all forms and sizes—if transformation is to occur at a rapid pace and at scale.

Farmers and the agri-food value chain need to produce more with less, which highlights the productivity and marketing challenges that are historically familiar. The opportunity to achieve these outcomes can only be realized if millions of smallholder farmers and—importantly—the "agripreneurs" (an entrepreneur whose main business is agriculture or agriculture-related) in the private sector are engaged. In addition to the impacts of climate change, there are other major issues in some developing countries. A triple crisis of hunger, micronutrient deficiency, and obesity exists: the double burden of malnutrition. The challenges facing the Asia and Pacific region and its subregions have increased by a magnitude. This means that investments need to be scaled up, all parts of the agri-food system—public and private sector—involved, and the capacity of developing countries strengthened to meet this increased threat.

Agriculture is both a victim of and a contributor to climate change. Global farming productivity has been suppressed by almost 21% in the last 60 years because of climate change. This trend will worsen in the coming decades and the agriculture and food system will be exposed to more lost productivity if reductions in greenhouse gas (GHG) emissions are not achieved. Global agriculture has grown more vulnerable and less resilient to ongoing climate change. Food security has become more uncertain and subject to greater risks.

Climate change negatively affects prices and revenues, as well as farm production and food security. Without effective interventions, scientists predict that adverse climate change effects will hit the incomes of vulnerable populations the hardest, and in coping with climate change risks, women particularly face special circumstances and higher risks than their male counterparts. Agriculture is affected by climate change everywhere with pervasive, multiple impacts. "Climate-smart" farming is a necessity.

The coronavirus disease (COVID-19) pandemic was the ultimate disrupter for the world agri-food system. It negatively affected and constrained consumer choices and demand, domestic supplies and foreign trade, food availability and nutrition outcomes for rural and urban populations, and the ongoing efforts to improve farming productivity and the rural environment. Adaptations to climate change (and other planned interventions) were interrupted and any that may have occurred were likely to be a matter of chance rather than because of a planned policy. Food security cannot be taken for granted and resilience has probably been undervalued in the past. Disruption also has a positive side: innovation (such as digitalization) is welcome because it offers ways to improve processes, increase productivity, meet the demands of consumers through the market, reduce emissions, and help to meet climate change targets. Climate-smart agriculture is often inclusive. It increases the participation of women in data gathering, decision-making, and access to agricultural resources.

Transforming agri-food systems requires innovation in technical solutions, and organizational and institutional changes. It's a complex challenge with many dimensions. Adopting a holistic view of the system is the minimum that is needed to address this complexity. Making connections across disciplines and ensuring synergy at scale can usually only be obtained by some form of automation or—at least—digitalization. These types of disruptive innovations can improve productivity on the farm and in the value chain and improve sustainability and nutrition security at the same time. Human and regulatory elements in agri-food systems may be obstructing digital innovations on the farm and in the value chain and related educational and health-care systems. There are case studies that show that successful applications and transformations can occur, but these lessons need to be applied more widely.

A caveat on digitalization is important: any bias in the accessibility of digital technology could lead to a bias in data collection and policy decisions. This is a danger that must be avoided. The potential gains, however, for improved public information systems (especially for designing development finance and aid, which are often remote from the market) are enormous.

The building blocks of natural capital—soil, water, air, biodiversity, and energy substrates—have been heavily exploited since the industrial revolution. These components of natural capital were often seen as public goods with low or zero protection from any authority. Their value was typically seen as zero since the market could set no prices for these "free goods"— a classic case of market failure. This has had negative consequences for the planet as successive feasibility exercises for investment were undertaken. Natural capital did not get a mention. An approach that aims to change this is the Natural Capital Project. The project has adopted the concept of gross ecosystem product and Integrated Valuation of Ecosystem Services and Tradeoffs (InVEST). InVEST is a group of free, open-source software models that map and value natural resources. This approach has the potential to change everything in the development community. Significantly—if adopted on a wide scale—climate change policies will improve living standards for developing countries and development policies will help mitigate climate change. The Natural Capital concept will shape the future of development economics, Asian Development Bank regional and national technical assistance, and credit finance operations for major infrastructure investments.

Nutrition security in the world is also alarming. According to the United Nations, the world is not expected to achieve targets for any of the major nutrition indicators by 2030. *The State of Food Security and Nutrition Report in the World 2021*, published by the Food and Agriculture Organization of the United Nations (FAO), states that states that more than half of the undernourished people in 2020 are from Asia. The FAO notes that "With less than a decade to 2030, the world is not on track to ending world hunger and malnutrition; and in the case of world hunger, we are moving in the wrong direction." The nutritional challenge in many developing and emerging economies is a double challenge: the double burden of malnutrition (DBM), infant and child undernutrition occurring at the same time as overnutrition and obesity in the population. Two out of every five adults are obese in the Asia and Pacific region. As of 2022, the region has the highest absolute number of overweight and obese people at 1 billion. At the same time, the percentage of stunted, wasting, and underweight children in South Asia is among the highest in the world.

DBM is the prelude to the rapidly growing burden of noncommunicable diseases (NCDs) that already account for 75% of deaths worldwide. Early life undernutrition—starting as early as in utero—not only predisposes children to poor physical and cognitive development in life but also an increased risk of NCDs in adulthood. Even without this impact on health and life opportunities, there are straightforward and easily justified economic reasons for addressing DBM. Poor health holds back the productivity and economic progress of a population, an impact that can be measured and for which interventions can show an economic payback. At a micro level, individuals and families bear the crushing medical costs of dealing with NCDs. On a macro level, health-care budgets will rise, and there is a loss of productivity and missed opportunities for the nation. These double duty actions comprise interventions and activities that can reduce the risk or burden of both undernutrition (including wasting, stunting, and micronutrient deficiency or insufficiency) and overweight, obesity, or diet-related NCDs at the same time. Policy instruments are the way forward, but they are complex to design, administer, and implement. To succeed, the approach of the policy maker and the development community must be a holistic one. Official departments and development partners should work together for coordinated thinking and action. A systems approach requires a shift away from fixed, fully planned programs to more iterative and adaptive planning, and a focus on co-creation with local stakeholders.

In almost every country, there is a gap between rural and urban dwellers, even within developed economies. This is often measured in terms of income and average living standards, but many other metrics can be used to report on the quality of life and opportunities for advancement for rural dwellers and for those who live in urban areas. In less developed and emerging economies, the gaps between rural and urban dwellers tend to affect a much higher proportion of the population than in developed economies, and deficiencies are likely to have a much larger impact on economic growth. They may also, locally and nationally, have an impact on climate change. Poor rural dwellers are more inclined to overstock their animals on pastures or cut down trees and forests, for example, to feed themselves. Poverty in rural areas is often characterized by local food insecurity and it contributes to national and global food insecurity. Climate change and protecting biodiversity are other reasons to reduce the rural–urban divide.

Human capital endowments often play a big part in decisions to migrate from rural areas in developing countries—or subregions—where economic opportunities in rural areas are restricted. Rural-urban migration decisions may be based on prospective access to schools and health care. The situation for Asia—and particularly the People's Republic of China (PRC) and its rural population—has been examined for data on educational standards, incomes, and employment. A review of the experience of secondary education development in the Republic of Korea emphasized the importance of universal secondary education, special policies and investment for rural areas, the investment in technical and vocational education and training high schools, and incentives for teachers who worked in rural areas. The PRC may be underinvesting in these types of interventions, and data from Bangladesh, India, and the PRC illustrate the ongoing divisions between urban and rural populations. Apart from these macro-level observations on investment in rural education and training, micro-level, innovative systems for growing food—such as vertical farming—are possible in peri-urban locations. Training can improve the understanding of food production and help improve food security at the same time. Education and training inputs combine to support the increased production of quality, sustainable food.

A global effort is now underway to redirect the economic and environmental activities of humanity. The development community (donor countries, agencies, banks, and partners) is playing its part in this redirection, as demonstrated by contributions at the RDFS Forum 2022. The community offered targets, resources, and practical ideas to improve economic and human development, enhance food security, and adapt to and mitigate climate change.

COVID-19 has reminded people and policy makers that pandemics are a risk factor that can throw government and development practitioner plans off course. While conflicts, war, and political upheaval may have been confined to some subregions and populations, they have been shown to have wider global impacts. Overall, business plans have become riskier. Risk and uncertainty are now increasingly important in the development environment, and they further emphasize the need for resilience and sustainability when faced with external shocks.

Challenges are growing and are increasingly complex in nature and impact. A common reaction to these challenges has been to recommend and design policies that improve human capital and work across several dimensions: health, food policy, education, social protection, the environment, etc. This observation helps form an initial conclusion about the way forward for climate change policies. The future for interventions that transform agriculture, improve food security, reverse climate change, and improve nutrition and health will be a holistic one. Multidisciplinary efforts and collaborative actions by governments and development partners are essential and must embrace a more coordinated—rather than a siloed—way of thinking.

COVID-19 revealed that market signals and private sector operators can be powerful forces in identifying issues and redeploying resources to meet demand and help modify and transform supply chains. Disruption can be a positive experience. The market is a source of ideas, resources, and skills, and offers inclusivity, subject to market participants having finance. The purchasing power of smallholders and consumers may be small at an individual level, but even relatively small amounts of producer-sourced finance or consumer-driven demand can leverage resources and change at a local level— especially if the market is innovative about how producers and consumers interact and act collectively.

Enabling and leveraging market forces and private sector operators in all forms and sizes needs to be seen as a desirable strategy to effect change and transformation. Facilitating actions by the development community may take the form of providing infrastructure (smartphones, Wi-Fi, broadband, etc.) and programs that deliver knowledge, training and education, and access to digital systems. In parallel, identifying and supporting agripreneurs and progressive private firms are aspects of human capacity development. Producers and economic actors are on the front line of development and can be enormously powerful in spreading ideas and innovation, especially if stakeholders and development partners offer support to achieve scale.

Resilience and risk are aspects of development that may have historically been underplayed. There have always been vulnerability to natural hazards—earthquakes, floods, droughts, crop failures, and livestock diseases—but the world now seems to be operating in a different, higher range of values for risk. Climate change is the main cause of extreme weather events, and disease outbreaks are exacerbated by the exploitation of natural resources, and population growth and urbanization. The increasing and climate-induced volatility of the world cannot be dealt with by smallholder farmers. The reaction of many to increased risk would be to cut back on investment and new forms of production. This means that development partners and stakeholders may need new substantive ways of dealing with risk and uncertainty in the future, as it will not be enough to leave this to "the market."

One aspect of risk that deserves separate mention is social protection. COVID-19 brought this issue into focus as finance ministers around the world realized that the pandemic and its associated lockdown policies had the potential to bring economic and social catastrophe, even in developed economies. The experience of COVID-19 is a reminder that an economic system will struggle if there are not enough people to maintain and manage it. Another positive aspect of the pandemic is the growing understanding that measures to improve social protection in certain circumstances has grown.

Green finance and the relatively new subject of natural capital are other areas of thinking and work that will profoundly change calculations about development activity and program design. The widespread adoption of the concepts of gross ecosystem product and InVEST will be transforming. These ideas will be a necessary condition for future decision-making by all stakeholders, both within and outside the development community.

Success stories from commercial practice and the interventions made by development partners are important. These case studies—reported from all parts of the Asia and Pacific region—offer an observation of valuable lessons from the field. They are invaluable pointers for how to transform agri-food systems and meet the challenges of climate change, nutrition, and the rural–urban divide.

I. Introduction

Food security has always been a crucial issue for humankind but it demands more attention at this point in human history. There is an existential threat to civilization from climate change which directly impacts agriculture and the availability of food. Global events—the coronavirus disease (COVID-19) pandemic, extreme weather episodes, and military conflicts—underline this observation. This has been described as a perfect storm in the global agri-food value chain, a storm driven by high energy prices, pandemic-related disruptions, and the ongoing economic and political conflict, and set against a backdrop of continuing extreme weather events happening all over the world. The consequence of this for consumers in economies of all levels is that food security is less certain than it has been in previous decades. Globally, aid agencies, government departments, think tanks, and development banks are alerting the world to this issue. It is in this context that the Asian Development Bank (ADB)—in collaboration with several development partners organized the Asia-Pacific Rural Development and Food Security Forum 2022, which enabled around 1,000 stakeholders and experts to be together in a virtual meeting place.

Public acknowledgment of this new situation of the global agri-food system may, paradoxically, have a positive impact. Policy makers everywhere are now united on one thing: they do not assume that food security is a given. The 26th Conference of the Parties in 2021 offered evidence of this realization and the first 6 months of 2022 reinforced the need for action to address the challenges facing developing economies and the world. From individual crisis events, an opportunity has arisen: a renewed focus by decision-makers on how sustainable, affordable, and nutritious food can be produced for the world's people. There will be debate on the precise design of policy interventions, but interventions are necessary and urgent. The 2022 forum on rural development and food security was timely and brought together a group of experts, researchers, policy makers, and practitioners from the public and private sectors to consider the challenges and threats to food security. Forum participants presented ideas, shared experiences, and discussed how the future of the agri-food system can be shaped to support human progress and development and achieve climate change, food security, and Sustainable Development Goals. The ideas and experiences presented and discussed explored many of the changes and innovations that the agri-food system of the region—and its associated policy and financial interventions— needs to adopt as it transforms itself to overcome threats to food security.

This forum is taking place at a time when the world is facing escalated food and nutrition risks, and the hunger and nutrition related achievements of the last 2 decades or so are showing reversal.
– Roberta Casali, ADB Vice-President for Finance and Risk Management

This report on the forum's program will proceed using four main themes:

- sustainability, digital technology, and inclusive food systems;
- financing sustainable agriculture;
- nutrition security and the double burden of malnutrition; and
- the rural–urban divide.

These themes are interrelated and are sometimes difficult to separate, but the forum brought together experts who could contribute significantly to each of these areas. This paper reports on the discussions and contributions during the proceedings.

The Asia and Pacific region is dominated by smallholders (tilling less than 2 hectares). However, many smallholders earn just one-third of what is a minimum livable income and their productivity is often well below the potential of the resources utilized. Smallholder farmers often struggle financially. The COVID-19 pandemic has made matters worse, but climate change has been increasingly evident and damaging through increasing average temperatures, lethal heat waves, extreme precipitation events and floods, severe hurricanes, drought, and changes in water supply. Added to this we have a war in Europe in 2022 that has the potential to increase hunger and famine for millions of people.

Asia and the Pacific is probably the region that is most exposed to physical climate risk in the world. It is frequently impacted, and unless adaptation and mitigation policies are adopted on a wide scale, Asia will inevitably experience severe consequences of climate change. In particular, countries in Asia with lower levels of per capita gross domestic product—for example, Bangladesh, Cambodia, India, Indonesia, the Lao People's Democratic Republic, Malaysia, Pakistan, the Philippines, Thailand, and Viet Nam—will fare the worst. Pacific islands—such as the Cook Islands, Fiji, the Marshall Islands, Nauru, Papua New Guinea, Samoa, Solomon Islands, Tonga, Tuvalu, and Vanuatu—are threatened by storm surges and rising sea levels. But all countries in the world will be affected by the impacts on domestic agricultural production or international trade in commodities. No country can isolate itself from climate change. By 2050, it is expected that countries in the Asia and Pacific region will experience considerable increases in heat and humidity, as well as a greater propensity for extreme precipitation and storms. As the physical thresholds that affect human beings (heat, cold, rain, wind, and humidity) are reached they will affect the ability to work outside, which implies that the agriculture and rural sectors will be hit the hardest. These countries also have limited resources—less accessible credit and finance, and less well-established systems—which are needed for adaptation.

Financing investment in the 21st century isn't as straightforward as in previous millennia. It is now recognized that investments must not just be feasible and economically viable. They also must be sustainable. This partly explains how the term "green finance" has entered the lexicon for discussions around sustainable agri-food systems. This is a positive development in principle, but the implementation of this idea is not straightforward. The forum gave prominence to the intellectual and practical aspects of this subject through a discussion—with case studies—of the natural capital approach to financing development opportunities.

The COVID-19 pandemic caused many problems and challenges within the agri-food value chain in the Asia and Pacific region. Agriculture and markets were severely disrupted by the virus and its related control measures. For some, crops and livestock yields were badly affected as labor became scarce (reduced incomes limited farmer ability to hire farm labor or its use was highly restricted because collective work was banned in fields or collection centers, or migrant labor went home). Due to quarantines, ill truck drivers, border closures, and trade restrictions, aggregators were unable to purchase at the farm gate, marketplaces were blocked, and supply chains were constrained. The pandemic gave women and girls extra burdens as family members became ill and less available for work, and sick families needed carers, the traditional role of women. However, in this pandemic-induced crisis of the food value chain, opportunities arose for "agripreneurs" (an entrepreneur whose main business is agriculture or agriculture-related) and the application of new systems using new technology. Direct connections with customers were identified and new systems were designed that reduced the number of actors in the value chain or increased their efficiency. Digital innovation enabled new opportunities to become realities. Farmers and traders at the beginning of the value chain were able to make rapid and efficient connections with buyers and consumers at the end of the value chain. The crisis of COVID-19 stimulated entrepreneurial talent and digital technology and encouraged farmers and traders to overcome dysfunctions in the markets and make direct sales to consumers in regions where previous attempts at regulatory reform and support programs had failed.

The vulnerability of the Asia and Pacific region to climate change is of concern, but there is another aspect of the food system in the region that needs to be addressed: nutrition. One of the biggest nutritional challenges in many developing and emerging economies is the double burden of malnutrition and linked diseases: infant and child undernutrition occurring at the same time as adult overnutrition and obesity. For example, the percentage of stunted, wasting, and underweight children in South Asia is among the highest in the world. This also applies to those who are micronutrient-deficient or those who lack iron, vitamin A, and zinc. At the same time, two out of every five adults are obese in the Asia and Pacific region. As of 2022, the region has the highest absolute number of overweight and obese people, at 1 billion.

Developing and emerging economies require strategies to improve their nutrition while avoiding having the "wrong food" in the system: food that is not nutritious, or which encourages overeating for adults and the consequent ill health associated with obesity. Globally in 2022, there is more obesity than there is under nutrition. In the developing world, obesity prevalence is catching up with undernutrition, while in the transitioning economies there is more obesity than there is undernutrition.

Sustainable rural development is an essential element of any attempt to create a sustainable agri-food system. Because rural populations have opportunities to move to urban areas (and migration occurs at high growth rates in almost all of Asia) the economic, social, and environmental balance of regions and countries is affected, which can then affect agricultural transformation. Poverty and a weakened stock of human capital in rural areas will create a rural–urban divide which is inimical to making efficient and timely changes to how producers and the value chain can adopt new ways of operating. These observations suggest that education, upskilling, and training initiatives in rural areas have an important part to play in addressing these divisions and the indirect effects of migration. Rural development strategies should include targeted, diverse methods and consider the human capital and educational capacity in these regions.

The forum program covered four main themes: (i) sustainable and inclusive food systems and the contribution of digital technology, (ii) financing sustainable agriculture and natural capital, (iii) nutrition security and the double burden of malnutrition, and (iv) the rural–urban divide. The content that the forum participants delivered is incorporated in this report. It conveys the ideas and experiences of the participants at the forum and identifies the key challenges and levers for change in a world that is threatened by climate change. The concluding section presents readers with solutions for the challenges that the developing world faces.

Program

Day 1: 22 March 2022	
2:00–2:02 p.m.	**Introduction** **Shiulie Ghosh**, Director, Aero Productions Ltd.
2:02–3:05 p.m.	**Leaders' Roundtable: The Future of Food and Agriculture** Climate change, the COVID-19 pandemic, shifts in dietary preferences, and economic and demographic transformations are among the factors influencing the future of food and agriculture. The session will discuss ways to rethink agri-food system transformation, while considering effective models of governance and collaboration, to meet the continuing demand for safe, nutritious, and affordable food. Global leaders will share their perspective on how to build a green, nature-positive, sustainable, and resilient food system. **Moderator: Shiulie Ghosh**, Director, Aero Productions Ltd. **Welcome address: Qingfeng Zhang**, Chief, Rural Development and Food Security (Agriculture) Thematic Group, concurrently OIC, Environment Thematic Group, Sustainable Development and Climate Change Department, Asian Development Bank **Opening Remarks** **Roberta Casali**, Vice-President (Finance and Risk Management), Asian Development Bank **Donal Brown**, Associate Vice-President, Programme Management Department, International Fund for Agricultural Development **Panel Discussion** 1. **Shenggen Fan**, Chair Professor, China Agricultural University 2. **Yutaka Arai**, Vice-Minister for International Affairs, Ministry of Agriculture, Forestry and Fisheries, Government of Japan 3. **Cao Đức Phát**, Vice Chair, Board of Trustees, International Rice Research Institute

3:05–3:15 p.m.	Session Evaluation and Break

Technical Session 1: Digital Technology for Agriculture

The COVID-19 pandemic promoted a surge in digital applications in agriculture, highlighting technology's potential to help modernize agriculture and transform food systems. The session will focus on the role of digital technologies in transforming food systems to offer increased profits for farmers and entrepreneurs in the value chain. Discussions highlight requirements to make digitalization of agriculture happen such as policy interventions, public-private partnerships, financing, infrastructure development, and capacity development, among others.

Moderator: Thomas Panella, Director, Environment, Natural Resources and Agriculture Division, East Asia Department, Asian Development Bank

Key Issues: precision agriculture, e-commerce, digital divide, inclusiveness, digital infrastructure, human capital, fiscal, monetary, and regulatory framework for digital agriculture

3:15–3:30 p.m.	**Keynote Address**
	Paul Teng Piang Siong, Managing Director & Dean, NIE International Pte Ltd | Nanyang Technological University, Singapore
3:30–4:05 p.m.	**Panel Discussion**
	Panelists
	1. **Vladimir Stankovic**, Program Coordinator, International Telecommunication Union
	2. **Elliott Grant**, General Manager, X
	3. **Vikas Choudhary**, Senior Economist, World Bank
4:05–4:40 p.m.	**Revisit and Reflect for Revision: Experience Sharing, Voices from Development Practitioners, Startups, Farmers, and Youth**
	1. **Don Tan**, Director-Corporate Affairs, Pinduoduo Inc.
	2. **Otini Mpinganjira**, Program Lead, PxD
	3. **Jawoo Koo**, Senior Research Fellow, International Food Policy Research Institute
	4. **Takeshi Ueda**, Principal Natural Resources and Agriculture Economist, Asian Development Bank
4:40–5:05 p.m.	**Q&A and Open Discussion**
	Session Evaluation
5:05–5:15 p.m.	**Synthesis: Day 1 Activities and Sessions**
	1. **Shingo Kimura**, Senior Natural Resources and Agriculture Specialist, Asian Development Bank
	2. **Navin K. Twarakavi**, Senior Digital Agriculture Specialist, Asian Development Bank

Day 2: 23 March 2022

11:35–11:40 a.m.	**Recap: Day 1 Activities and Sessions**
	Md Abul Basher, Senior Natural Resources and Agriculture Specialist, Asian Development Bank

Deep Dive 1: The Role of Education in Reducing Rural–Urban Divide

To avoid the middle-income trap, countries must move from a low-cost to a high-value economy. This development, however, has the potential to create disparities including urban-rural inequality. Reducing the rural–urban divide is critical for more inclusive growth and necessary for sustainable economic growth. Discussions highlight how a rural–urban integrated development can assume a multidimensional focus that is geared toward delivering well-being and prosperity to rural dwellers which is comparable to that in urban areas.

Moderator: Brajesh Panth, Chief, Education Thematic Group, Asian Development Bank

Key Issues: rural and urban income disparity, integrated rural development, human capital investment, rural economic hub, and social and economic welfare

11:40–11:45 a.m.	**Welcome Remarks** **Qingfeng Zhang**, Chief, Rural Development and Food Security (Agriculture) Thematic Group, concurrently OIC, Environment Thematic Group, Sustainable Development and Climate Change Department, Asian Development Bank
11:45 a.m.–12:00 noon	**Keynote Address** **Scott Rozelle**, Helen F. Farnsworth Senior Fellow and the co-director of the Rural Education Action Program, Freeman Spogli Institute for International Studies, Stanford University
12:00 noon–12:30 p.m.	**Panel Discussion** **Panelists** 1. **Clarissa Delgado**, CEO, Teach for Philippines 2. **Md. Saidur Rahman**, Professor & Former Head Dept. of Agricultural Economics, Bangladesh Agricultural University 3. **Meekyung Shin**, Education Specialist, Asian Development Bank
12:30–12:55 p.m.	**Q&A and Open Discussion**
12:55–1:10 p.m.	Session Evaluation and Break

Deep Dive 2: Urban/Controlled Environment Farming—New Window for Fresh and Nutritious Food

Urbanization has been rapidly increasing in Asia and the Pacific over the last 2 decades. With more people now living in cities, the demand for food has become greater in urban areas. Urban farming presents an increasingly viable method of not only enhancing urban food security but also reducing the pressure on traditional agricultural land thereby contributing to a reduction in carbon footprint.

Discussions focus on the growth of urban farming as a viable method to produce food, and how new farming approaches, appropriate urban design, and government policies can help propagate this concept.

Moderator: Michiko Katagami, Principal Natural Resources and Agriculture Specialist, Asian Development Bank

Key Issues: increasing urbanization, new farming approaches (vertical farming), controlled-environment agriculture, appropriate urban design (e.g., community gardens, urban wastewater for irrigation, organic waste as compost, etc.), impact on food waste, food security in island nations

1:10–1:25 p.m.	**Keynote Address** **William Dar**, Secretary, Department of Agriculture, Philippines
1:25–2:25 p.m.	**Panel Discussion**

Panelists
1. **Sairam Reddy Palicherla**, Co-Founder and Chief Scientific Officer, UrbanKisaan Farms Pvt Ltd.
2. **Maria Tran**, Senior Project Officer (Urban Development), Asian Development Bank
3. **Gerald Glenn Panganiban**, National Program Director for Urban Agriculture at Department of Agriculture - Philippines

Q&A and Open Discussion

2:25–2:40 p.m.	**Session Evaluation and Break**

Technical Session 2: Pathways to Sustainable and Inclusive Food Systems

Climate change and resource-intensive agricultural practices are causing stress and risks to the environment, affecting food production and its value chains. The session will emphasize the benefits of climate-smart approaches and practices across crop, fish, and livestock production considering nature- based solutions and the One Health approach. Discussions will focus on innovations, on-the-ground experiences, and lessons learned, including feminization of agriculture and youth empowerment.

Moderator: Jiangfeng Zhang, Director, Environment, Natural Resources and Agriculture Division, Southeast Asia Department, Asian Development Bank

Key Issues: climate-smart agriculture, incentivization of climate-friendly production and consumption, fiscal, monetary, and regulatory framework for climate-smart agriculture (crop, fish, and livestock subsectors), One Health, and planetary health

2:40–2:50 p.m.	**Opening Remarks** **Bruno Carrasco**, Director General, concurrently Chief Compliance Officer, Sustainable Development and Climate Change Department, Asian Development Bank

Keynote Address
Louis Verchot, Principal Scientist and Leader of the Land Restoration Group, International Center for Tropical Agriculture

3:05–3:35 p.m.	**Panel Discussion**

Panelists
1. **Jean Balie**, Director General, International Rice Research Institute
2. **Sudarshan Dutta**, Lead Agronomist, Agoro Carbon Alliance (Yara)
3. **Samantha Hung**, Chief of Gender Equality Thematic Group, Asian Development Bank

3:35–4:00 p.m.	**Revisit and Reflect for Revision: Sharing of Experiences, Voices from Development Practitioners, Farmers, Youth, and Women** 1. **Omer A. Zafar**, Principal Natural Resources and Agriculture Specialist, Asian Development Bank 2. **Romina Cavatassi**, Lead Economist, International Fund for Agricultural Development 3. **Le Hoang Anh**, Senior Official, Department of Science Technology and Environment, Ministry of Agriculture and Rural Development, Viet Nam
4:00–4:30 p.m.	**Q&A and Open Discussion**
4:30–4:40 p.m.	**Session Evaluation and Break**

Technical Session 3: Intersectoral Approach to Nutrition Security

One size does not fit all when addressing the double burden of malnutrition. Cross-sectoral approaches are needed but difficult to implement. This session will present on-the-ground experiences, lessons learned, and case studies on best practices to collectively work together on what can be done in terms of policy interventions, regulations, financing, food production innovations, and others to promote nutrition security.

Moderator: Yasmin Siddiqi, Director, Environment, Natural Resources & Agriculture Division, Central and West Asia Department, Asian Development Bank

Key Issues: double burden of malnutrition, undernutrition, obesity integrated solutions addressing the double burden of undernutrition and obesity in developing Asia

4:40–4:55 p.m.	**Keynote Address** **Mandana Arabi**, Vice President, Nutrition International
4:55–5:30 p.m.	**Panel Discussion** **Panelists** 1. **Ladda Mo-suwan**, Professor, Department of Pediatrics, Faculty of Medicine, Prince of Songkla University 2. **Britta Schumacher**, Senior Nutritionist, Regional Bureau for Asia-Pacific (Bangkok), World Food Programme 3. **Victor Ochieng Owino**, Nutrition Specialist, Division of Human Health, International Atomic Energy Agency 4. **Isaac Kofi Bimpong**, Plant Breeder / Geneticist, Department of Nuclear Sciences and Applications
5:30–6:00 p.m.	**Revisit and Reflect for Revision: Sharing of Experiences, Voices from Development Practitioners, Consumers, and Women** 1. **Kaz Maruyama**, President and Representative Director, DSM Japan 2. **Jody Harris**, Lead Expert, World Vegetable Center 3. **Ahmed Umair**, Chief Executive Officer, Vital Agri Nutrients Ltd, Pakistan
6:00–6:30 p.m.	Q&A and Open Discussion

6:30–6:40 p.m.	Synthesis: Day 2 Activities and Sessions
	1. **Narayan Iyer**, Senior Natural Resources and Agriculture Specialist (Agribusiness), Asian Development Bank 2. **Kazuko Ogasahara**, Senior Natural Resources and Agriculture Specialist (Health and Nutrition), Asian Development Bank
	Session Evaluation

Day 3: 24 March 2022

11:30–11:35 a.m.	Recap: Day 2 Activities and Sessions
	Sangjun Lee, Natural Resources and Agriculture Specialist, Asian Development Bank
11:35 a.m.–12:35 p.m.	Special Session: COVID-19 Impacts on Food Systems
	Moderator: **Takashi Yamano**, Principal Economist, Asian Development Bank
	Keynote Address **Aziz Elbehri**, Senior Economist and Stream Leader, Agri-food, Rural Development and Socio-economic Policies, Food and Agriculture Organization Regional Office for Asia and the Pacific
	Panelists 1. **Yasmin Siddiqi**, Director, Central and West Asia Department, Asian Development Bank 2. **Jalil Pirzada**, Director, Institute of the Economy, Analysis and Agriculture Development, Ministry of Agriculture, Tajikistan 3. **PK Joshi**, Former Head, International Food Policy Research Institute
12:35–12:50 p.m.	Break

Deep Dive 3: Alternative Proteins to Meet the Growing Demand

With the pace of global population growth and improvement of dietary habits in emerging countries, protein supply requirements in 2050 will be twice as much as they were in 2005. As a result, the supply of protein will be insufficient as early as 2030. Alternative protein sources— such as plant-based, microbial- based, and edible insects— are "eco-friendly" protein sources, offering potential solutions to the global problem of food shortage, over-farming, and depletion of natural resources. Discussions highlight issues related to expanding the consumption of alternative protein sources including consumer acceptance, food allergies, nutritional balance, cost and availability, and relevant regulations.

Moderator: **Kate Jarvis**, Climate Finance Specialist, Agribusiness Investment Team, Private Sector Operations Department, Asian Development Bank

Key Issues: GHG emissions, deforestation, water intensity, healthy alternatives, advances in agrifood tech and synthetic biology, price-parity, government initiatives /accelerators, regulatory guidelines

12:50–1:05 p.m.	**Keynote Address** **William Chen**, Director, Food Science and Technology Programme, Nanyang Technological University
1:05–2:05 p.m.	**Panel Discussion** **Panelists** 1. **Varun Deshpande**, Managing Director, Good Food Institute 2. **Neil Ian Lumanlan**, Circular Bioeconomy Consultant 3. **Isabelle Decitre**, Founder and CEO, ID Capital **Q&A and Open Discussion** **Closing Remarks** **Martin Lemoine**, Principal Investment Specialist, Asian Development Bank
2:05–2:20 p.m.	Session Evaluation and Break

Technical Session 4: Financing Green, Resilient and Inclusive Agriculture

Financing is key to promote the implementation and adoption of innovations and technologies on-the- ground. This session discusses innovative approaches to mobilizing finances for smallholder farmers while emphasizing the role of multilateral development banks, public development banks, blended financing, and community-led financing, among others. Discussions highlight natural capital financing, and knowledge sharing on green valuation, eco-compensation, and digitizing supply chains.

Moderator: Mio Oka, Director, Environment, Natural Resources and Agriculture Division, South Asia Department, Asian Development Bank

Key Issues: innovative financing tools and approaches, cross-country and international experiences, role of multilateral development banks and private sector, blended financing, community-led financing, green financing, innovative natural capital financing facility

2:20–2:35 p.m.	**Keynote Address** **Gretchen Daily**, Bing Professor of Environmental Science, Stanford University
2:35–3:10 p.m.	**Panel Discussion** **Panelists** 1. **Qingfeng Zhang**, Chief, Rural Development and Food Security (Agriculture) Thematic Group, concurrently OIC, Environment Thematic Group, Sustainable Development and Climate Change Department, Asian Development Bank 2. **Piet van Asten**, Head Sustainable Production Systems – Coffee, Olam Food Ingredients 3. **Joost Zuidberg**, Management Board, AGRI3 Fund

3:10–3:40 p.m.	**Reflect and Rethink for Revision: Experience Sharing, Voices from Development Practitioners, Farmers, Youth, and Women** 1. **Kisa Mfalila**, Lead Environment and Climate Specialist-Asia and the Pacific Region, International Fund for Agricultural Development 2. **Stephen Hart**, Senior Loan Officer, European Investment Bank 3. **Arnaud Heckmann**, Principal Urban Development Specialist, Asian Development Bank
3:40–4:10 p.m.	**Q&A and Open Discussion**
4:10–4:25 p.m.	**Session Evaluation and Break**
4:25–4:55 p.m.	**Knowledge Product Showcase** Moderator: **Sungsup Ra**, Chief Sector Officer, Sustainable Development and Climate Change Department, Asian Development Bank Knowledge Product Showcase 1. **Shingo Kimura**, Senior Natural Resources and Agriculture Specialist, Asian Development Bank; Name of the report: "Financing Sustainable and Resilient Food Systems in Asia and the Pacific". 2. **Takeshi Ueda**, Principal Natural Resources and Agriculture Economist, Asian Development Bank; Name of the report: "Cambodia Agriculture, Natural Resources, and Rural Development Sector Assessment, Strategy, and Road Map"
4:55–5:05 p.m.	Break
5:05–5:50 p.m.	**Actions and Recommendations** Moderator: **Qingfeng Zhang,** Chief, Rural Development and Food Security (Agriculture) Thematic Group, concurrently OIC, Environment Thematic Group, Sustainable Development and Climate Change Department, Asian Development Bank Panelists 1. **Yasmin Siddiqi**, Director, Central and West Asia Department, Asian Development Bank 2. **Thomas Panella**, Director, East Asia Department, Asian Development Bank 3. **Mukhtor Khamudkhanov**, Director, Pacific Department, Asian Development Bank 4. **Martin Lemoine**, Principal Investment Specialist, Asian Development Bank 5. **Mio Oka**, Director, South Asia Department, Asian Development Bank 6. **Jiangfeng Zhang**, Director, Southeast Asia Department, Asian Development Bank 7. **Aziz Elbehri**, Senior Economist and Stream Leader, Food and Agriculture Organization of the United Nations
5:50 – 6:05 p.m.	**Closing Remarks and Future Direction** **Sungsup Ra**, Chief Sector Officer, Sustainable Development and Climate Change Department, Asian Development Bank
6:05–6:15 p.m.	**Credits, Note of Appreciation, and Forum Evaluation**

II. Sustainable and Inclusive Food Systems

A. Context

The productivity challenge for the world is now an even bigger challenge because climate change is working against the productivity gains derived from science and technology that are a key part of development for the agri-food value chain. Climate change significantly reduces the productivity of the agri-food system through its many impacts on weather, water availability, above average temperatures, etc. If climate smart policies are to be successful, agriculture—and its associated processing and distribution activities—must change. The agriculture and food system in 2022 is producing about one-quarter of total world greenhouse gas (GHG) emissions and if there is no reduction in food system-driven climate emissions, the world will not meet its 1.5°C target. Notwithstanding the impacts of emissions on temperature rise, land and natural resource degradation is widespread globally, and agriculture is responsible for some of this. The Food and Agriculture Organization of the United Nations (FAO) estimates that 34% of agricultural lands are degraded, 70%–80% of all forests worldwide have been altered, and South Asia and Southeast Asia contain many of the hot spots that exhibit these negative features.[1]

Food systems in Asia and the Pacific are highly vulnerable to climate change impacts but producers, processors, and consumers need to speed up adaptation to climate change and give priority to the transformation of the agri-food system. This will benefit all parties. Adaptive policies are a technically feasible and necessary step to achieve sustainable food and nutrition security, but governments, development banks, and their partners must participate so that the world moves from "theoretically feasible" to "field friendly." Private producers must also be involved—in all forms and sizes—if transformation is to occur at a rapid pace.

This report demonstrates what future discussions and decisions on agricultural and rural development should look like. An extraordinary transformation of the global agriculture and food system is urgently required to meet the future demand for sustainable food. Although it would be correct to describe the situation as a crisis, it is also an opportunity and one that forum participants recognized. The agri-food value chain can increase its productivity and reduce its emissions by investing in climate-smart agriculture that improves the resource efficiency of agricultural producers and enhances the resilience of food production and

---

[1] Source: Food and Agriculture Organization of the United Nations (FAO). 2021. The state of the world's land and water resources for food and agriculture – Systems at breaking point. Synthesis report 2021. Rome. Quoted in ADB. 2022. Rural Development and Food Security Forum 2022. Manila. https://doi.org/10.4060/cb7654en.

distribution systems. Farmers and the agri-food value chain need to produce more with less, which highlights existing productivity and marketing challenges. The opportunity to achieve these outcomes can only be realized, however, if millions of smallholder farmers and—importantly—agripreneurs are engaged in the private sector.

Environmental scientists and experts have identified, described, and measured climate change processes and effects, and the application of science will mitigate and speed up adaptation to climate change. But for those involved in international development, it is the job of policy makers, economists, social scientists, administrators and financiers, engineers, and entrepreneurs to take the next step: design and implement changes in the production, transportation, and consumption of food. Producers and economic actors in the value chain (input suppliers and postharvest agents) are on the front line. They must deal with climate change impacts each day and face the increasing demographic challenges of rural–urban migration, the feminization of agriculture, and aging farming rural communities. They must make a profit, pay back investment loans, and keep abreast of the changes that science and economics consider to be necessary. They should be included in the design and implementation of the transformation process.

With the coexistence of triple crises in some ADB developing member countries—hunger, micronutrient deficiency, and obesity—the region and its subregions face huge increases in the scale of the challenges. Investments need to be scaled up, all parts of the agri-food system involved—public and private,—and developing member capacity strengthened. Even in 2022, many dimensions of climate change are considered overwhelming, but they can be successful. Challenges need to be summarized and prioritized and, if possible, solutions provided through world-class research and best field practices. And if "best" practice cannot be agreed upon, multiple ideas and options should be quickly considered, using multiple situations.

B. Emissions, Climate Change, and Productivity Nexus

Agriculture is both a victim of and a contributor to climate change. A 2021 academic study underlined the huge impact of climate change on agriculture: it estimated that global farming productivity has been suppressed by almost 21% in the last 60 years because of climate change.[2] This is equivalent to losing roughly 7 years of agriculture sector growth since the 1960s. That growth has been instrumental in feeding the global population in the past and cannot be lost. This trend will worsen in the coming decades and the agriculture and food system will be exposed to more lost productivity if reductions in GHG emissions are not achieved. In addition to absolute losses, global agriculture has grown more vulnerable and less resilient to ongoing climate change. Food security has become more uncertain and subject to greater risks.

---

[2]   Source: A. Ortiz-Bobea et al. 2021. Anthropogenic climate change has slowed global agricultural productivity growth. *Nature Climate Change.* 11. pp. 306-312. Quoted in ADB. 2022. *Rural Development and Food Security Forum 2022.* Manila.

Systems of agricultural production in 2022 are contributing to emissions and have negative effects on farming and the food system through climate change impacts. As of 2022, Asia and the Pacific use around 65% of the world's water supply, and about 80% of fresh water in Asia and the Pacific is withdrawn for irrigation. However, irrigation efficiencies remain low, averaging 37%. In India, the efficiency of surface water and groundwater irrigation is at least 15%–20% below its potential. A water resource modeling exercise in 2017 has demonstrated that by 2050, 20% of the geographic area of Asia—which has a population of 1.6 billion–2 billion—may experience severe water scarcity.[3] Surface and groundwater sources in South Asia are likely to be under the greatest stress from increases in demand from agriculture, industry, domestic use, and climate change. Agriculture land and soil resources are being affected mainly due to either unsustainable agriculture intensification or poor farm management. In South Asia, about 43% of total agricultural land is degraded, with 31 million hectares already highly degraded. Farmland degradation in the region is characterized by soil erosion (water and wind), chemical deterioration including heavy metal contamination, physical deterioration including loss of organic matter and soil biodiversity, waterlogging, and aridification driven partly by climate change.

The growth in global emissions is, in part, caused by increases in the consumption of meat and dairy products in Asia, and the land reclamation and cultivation that is linked to unplanned agricultural development. During 1980–2020, meat consumption nearly tripled, and during 2010–2020 it increased by more than 30%. Dairy consumption increased by 70% during 1980–2020. The greatest consumption increases—greater than 3% per year (up until 2020)—have been seen in East Asia and Southeast Asia. Another contributor to harmful emissions is rice cultivation. It is estimated that rice cultivation contributes about 10% of anthropogenic methane (CH_4). Over 8.5 million hectares have been drained for agriculture in Southeast Asia, causing around 0.2 billion tons of carbon dioxide equivalent being emitted annually. Production of fish and shellfish in aquaculture in the region is calculated to be greater than 55 million tons (or about half of the global fish consumption). This production has high nitrous oxide (N_2O) emissions, which—by 2030—are expected to account for around a 6% increase in anthropogenic N_2O emissions. (Table). High rates of nitrogen fertilizer application drive high N_2O emissions in East Asia and South Asia.

Farm production and food security are also affected by climate change as there are negative effects on prices and revenues. Without effective interventions, scientists predict that adverse climate change effects will hit the incomes of vulnerable populations the hardest, and in coping with climate change risks, women particularly face special circumstances and higher risks than their male counterparts. Climate change affects agriculture everywhere, with pervasive, multiple impacts. "Climate-smart" farming is a necessity.

The organizational and institutional changes that are needed to apply pioneering, innovative technical, and other solutions in the agri-food system are complex.

---

[3]   Source: Y. Satoh et al. 2017. Multi-model and multi-scenario assessments of Asian water futures: The Water Futures and Solutions (WFaS) initiative. *Earth's Future*. 5 (7): pp. 832–852. https://doi.org/10.1002/2016EF000503.

Table: Net Anthropogenic Emissions Due to the Global Food System

Food system component	Emission (Gt CO_2 y^{-1})	Share of mean total emissions %
Crop and livestock production (N_2O and CH_4)	6.2 ± 0.3	12–13%
Deforestation and peatland degradation for food production (primarily CO_2)	4.8 ± 2.4	5–14%
Supply chain (primarily CO_2)	3.8 ± 1.3	5–10%
Food system total	14.8 ± 3.4	23–25%

CO_2 = carbon dioxide; CH_4 = methane; $GtCO_2y-1$ = billion metric tons of carbon dioxide per year; N_2O = Nitrous oxide

Source: L. Verchot. 2022. *Technical Session 2: Pathways to Sustainable and Inclusive Food Systems.* Quoted in ADB. 2022. *Rural Development and Food Security Forum 2022.* Manila.

Adopting a holistic view of the system is the minimum that is needed. Some development partners are restructuring or presenting detailed ideas on how development ambitions are attained.

C. Impact of COVID-19

Developing and emerging economies in the Asia and Pacific region faced the dual challenges of responding to the health crisis that the COVID-19 pandemic created while mitigating widespread economic disruption for their vulnerable populations. Even in developed economies, governments struggled to obtain the right balance between policies to maintain health and policies to maintain economic activity. Forum participants were not surprised to hear from Aziz Elbehri—from the regional office of the FAO—that the Asia and Pacific region was hit badly by the pandemic and that many countries are still suffering, with some experiencing a fifth wave of the virus. His contribution to the forum drew upon a review (produced by the FAO regional office earlier in 2022) of how the key indicators of economic activity had been affected by the pandemic.[4]

The pandemic affected virtually every part of the agri-food system, and there were wider effects on the macroeconomic system, consumer incomes, and nutrition. Some subsectors of the food system were almost wiped out (restaurants, tourist hotels) in some areas. Tourism, transport, and food service activities were severely curtailed while the demand for home delivery, online sales, and direct marketing soared. With an average gross domestic product decline of 4.1%, 2020 was essentially a lost year for the region, nullifying the regional 4% average growth rate in 2019.

---

[4] Source: A. Elbehri et al. 2022. COVID-19 pandemic impacts on Asia and the Pacific – A regional review of socioeconomic, agrifood and nutrition impacts and policy responses. Bangkok: FAO. https://doi.org/10.4060/cb8594en

Pandemic-induced lockdowns and restrictions badly affected employment. Debt levels increased for almost all sections of the population. Governments realized that social protection measures had to be implemented and paid for. The macro statistics show uneven impacts of the pandemic across urban and rural areas: urban areas reported more income loss and less economic activity, while rural areas saw greater food insecurity and increased poverty (relative to the urban areas). The absolute number of those defined to be in poverty may well have doubled under COVID-19. The number of undernourished people in the region rose from 322 million in 2019 to 376 million in 2020 (17%). Southern Asia saw the biggest increase in undernourishment in 2020 (Figure 1). There were also more pronounced impacts on women—notably more job losses, worse health impacts, and more domestic stress for females. Migration contracted in 2020, with negative impacts on migrant incomes and remittances in receiving countries.

The impact on consumer diets from COVID-19 has received relatively little attention but the forum heard that they were significant, with negative impacts from COVID-19 on nutrition in the Asia and Pacific region. In many Asian countries, consumers either reduced their consumption of nutritious food and/or turned to cheaper calorie-rich food. A reaction to lower incomes and unemployment was to cut the size of meals and reduce the intake of dense food. Urban households were reported as resorting to overeating and consuming low nutrition, high energy comfort food to cope with the stress:

- their intake of salty snacks increased;
- the intake of alcohol and sugary drinks increased substantially; and
- there was a greater dependence on online food delivery services, with a noticeable shift toward processed and imbalanced diets.

Figure 1: Impact on Consumer Diets from COVID-19

COVID-19: Food Insecurity Worsened Measurably

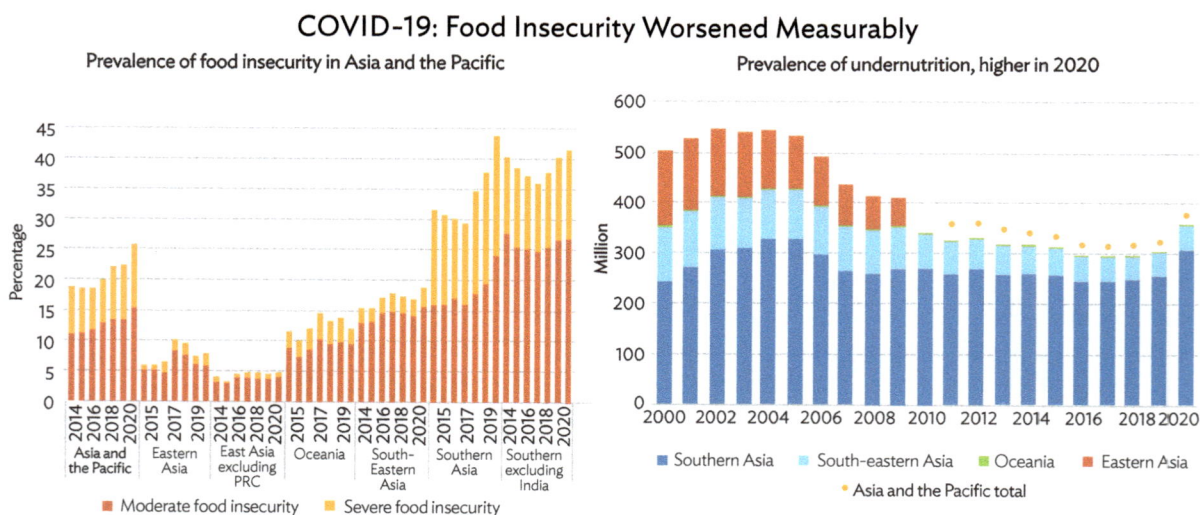

COVID-19 = coronavirus disease, PRC = People's Republic of China.

Source: A. Elbehri et al. 2022. *COVID-19 pandemic impacts on Asia and the Pacific – A regional review of socioeconomic, agrifood and nutrition impacts and policy responses.* Bangkok, FAO. https://doi.org/10.4060/cb8594en. Quoted in ADB. 2022. *Rural Development and Food Security Forum 2022.* Manila.

Overall, the pandemic was the ultimate disrupter for the agri-food system. It negatively affected and constrained consumer choices and demand, domestic supplies and foreign trade, food availability and nutrition outcomes for rural and urban populations, and the ongoing efforts to improve farming productivity and the rural environment. Adaptations to climate change (and other planned interventions) were interrupted and any that may have occurred were likely to be a matter of luck rather than a result of a planned policy. Even with exemptions to quarantine and lockdown policies and belated vaccination policies, there were often shortages of farm and rural labor and the supply of farming inputs, which affected productivity. Carrying out postharvest transport and processing operations was more difficult and sometimes impossible. Perishable crops and animal products were sometimes thrown away for lack of marketing capacity and/or people. And almost everywhere the demand for food changed radically as quarantine and lockdown policies hit consumer spending power and the normal patterns of work and leisure. Many of these damaging impacts on developing economies have occurred at various points in history but never all of them at the same time and in so many countries simultaneously.

Discussion has emphasized the negative impacts of climate change and the COVID-19 pandemic. The forum, however, presented several examples of how the disruptive tendencies of the pandemic encouraged innovation and led to positive impacts on the agri-food system. During the COVID-19 emergency, there were new opportunities that arose because of the pandemic. Investments in digital technologies to improve linkages with the final market and e-commerce trade were suddenly the obvious (and only) way to connect producers and consumers when traditional logistics, intermediaries, and supply chains were inhibited by quarantine and lockdown. In some countries services to rural communities, rural villages, and the rural economy were also boosted by the more rapid uptake of digital systems.

D. Women in Food Systems

Discussion of transformation in the agri-food systems should consider how gender inequality and barriers to the involvement of women can hinder progress in this area. Social and cultural gender norms can further constrain access by girls and women to health, education, training, jobs, financing, and mobility, contributing to unequal access by women to resources. There is evidence that demonstrates how climate change is disproportionately affecting women. Due to limited entitlements, assets, and access to the social and natural resources needed for adaptation and resilience building, women farmers are disproportionately affected by climate variability and weather extremes. Extreme climate events and climate-related disasters have often resulted in women and girls taking up additional duties as a laborer and placed additional burdens on their role as caregivers. Male out-migration has been a feature of many rural areas, and females are often the majority among smallholder farmers. In developing nations, 80% of working women are involved in food production. Although climate change may worsen existing gender inequalities in agriculture, it may also present new opportunities to realize the potential of women as agents of change and resilience building in the agri-food value chain. New opportunities are best realized via climate-smart agriculture (CSA) technologies and systems. Given that women and girls are often the majority of farmers, any program of knowledge transfer and upskilling needs to identify interventions that will directly benefit women and reduce gender gaps. A gender-responsive approach to CSA recognizes the capabilities of men and women to overcome barriers that they experience. These CSA techniques will deliver environmental benefits and will usually also reduce the labor burden for women in agriculture e.g., seeded rice, zero tillage machines, laser land leveling, green manuring, crop harvesters, weeding, solar pump irrigation, and postharvest management practices.

It is critical to recognize the role of climate smart agriculture in enhancing the access of women to agricultural resources and decision making. This approach can also provide linkages to new market opportunities. But there are structural issues to address. Women may have limited access to credit, extension services, and knowledge products. They may have restricted membership in cooperatives and water user associations, and they may have limited access to land and their right to land may not be recognized. Transformation to sustainable agri-food systems needs to support women to be agents of change and reform the discriminatory and sociocultural practices that limit their full participation in the agriculture sector.

E. Science That Improves Productivity and Reduces Emissions

Modern science, new technologies, and best practice case studies in the field offer many solutions to the challenges that low productivity and climate change bring. For example, FAO data show that while developing economies contain around three-quarters of the global cattle or buffalo population, they account for only one-third of global livestock production (meat and dairy). Farmers in developing economies tend to increase the size of their herds or flocks to increase milk and meat production, rather than increasing production per animal. The result is that units of meat or milk produced in developing economies will have higher levels of CH_4 emissions. They are very inefficient in producing meat and milk per unit of input and very efficient at producing methane. Thus, increased demand for milk and meat in developing economies increases methane emissions from domestic livestock. This situation must be mitigated and reversed and scientists and technologists offer many ideas on how to do this. The productivity of crops and livestock can be improved by improving genetics, applying modern husbandry techniques, and applying innovations to harvesting, feeding, and waste reduction systems that reduce CH_4 emissions. Digitalization of agriculture, biotechnology, novel farming environments, production systems, robotics, and novel food (alternative proteins) are just some of the areas that can help farmers and value chain actors to deliver more from less. Most—if not all—of these solutions are "shovel ready." All that is needed is investment finance, and governance and regulatory reform coupled with training to improve "human capacity."

Modernizing smallholder agriculture will improve outputs for the same number of inputs. This is a necessary improvement in technical productivity and is relevant to developing economies where employment opportunities in urban areas are growing. But this is not sufficient. Implementing key performance indicators—such as innovative animal and feed management techniques that target emissions—will improve many ruminant livestock production systems. CH4 inhibitors, oils and fats, oilseeds, electron sinks, and tanniferous forages will produce absolute reductions in emissions of around 20% with no negative effect on live weight gain. Diets can also be reformulated in ways that reduce dietary forage-to-concentrate ratios, increase feeding levels, and decrease grass maturity, which will reduce emissions. There are also innovations and climate smart improvements for crop production. In rice farming, breeding new varieties will support resilience to climate shocks and water, nutrient, and residue management can reduce CH4 emissions by about 80%, although this may increase N2O, so the net loss in GHG emissions is about 50%.

F. The Digital Revolution

There are increasing numbers of practical examples of how the digitalization of the agri-food system is delivering results. The application of digital technology is no longer just a theoretical concept, and these examples offer insights into how digitalization can disrupt agriculture and the food system (Figure 2). Disruptive innovations are seen by several commentators as potential saviors of the ecosystem in 2022. Disruptive innovations can improve productivity on the farm and in the value chain and improve sustainability and nutrition security at the same time.

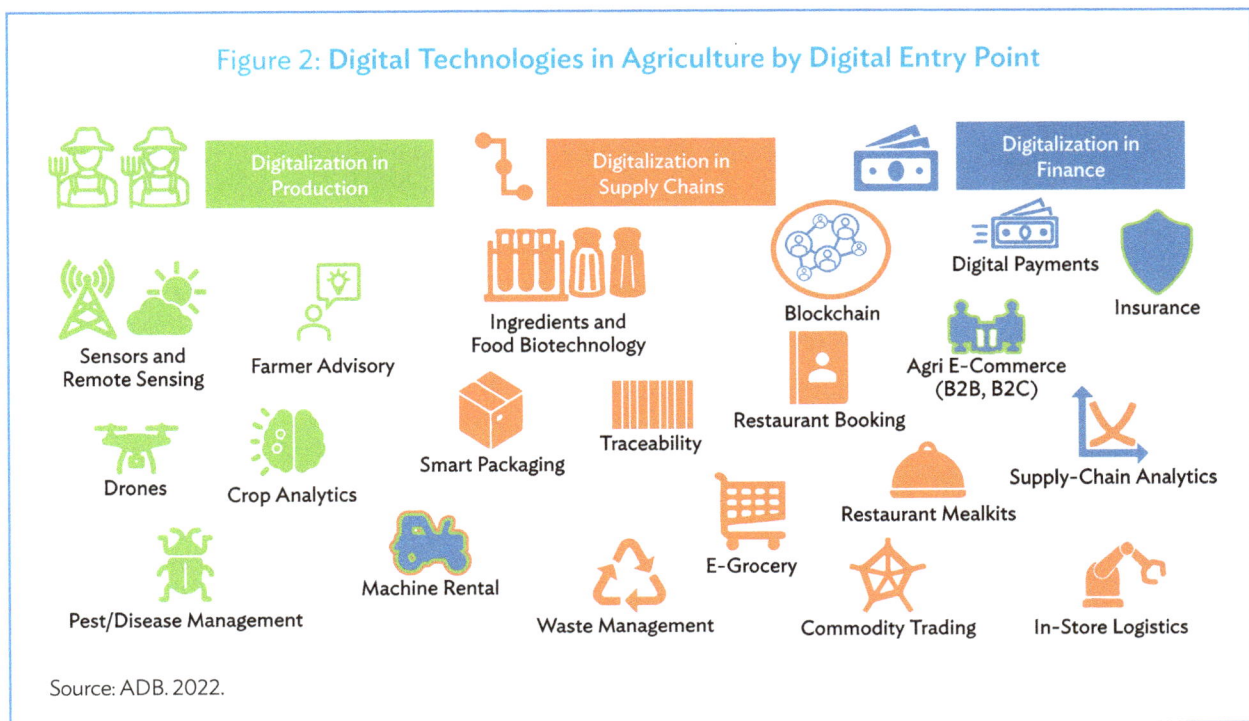

Figure 2: **Digital Technologies in Agriculture by Digital Entry Point**

Digitalization in Production
Digitalization in Supply Chains
Digitalization in Finance

Sensors and Remote Sensing
Farmer Advisory
Ingredients and Food Biotechnology
Blockchain
Digital Payments
Insurance

Drones
Crop Analytics
Smart Packaging
Traceability
Restaurant Booking
Agri E-Commerce (B2B, B2C)
Supply-Chain Analytics

Pest/Disease Management
Machine Rental
Waste Management
E-Grocery
Restaurant Mealkits
Commodity Trading
In-Store Logistics

Source: ADB. 2022.

Listing these entry points is useful—although not definitive—as more uses and applications for digitalization occur almost daily. Decision makers need timely, reliable, and actionable information and digital technology has all these attributes, and at lower costs than existing systems.

Decision points in the agri-food system where digital technology can impact:

- Biotechnology for crop and animal improvement, including nutrition and emissions;
- Novel environments for farming;
- Product integrity and fraud prevention;
- Supply chain logistics, infrastructure, and risk management;
- Novel food (e.g., alternative proteins); and
- Waste reduction and waste valorization.

Two major impacts of digital technology are in areas where policy makers and practitioners have recognized the need for reform for many years: providing low cost, timely advisory inputs for producers, and connecting producers with final markets and consumers. These service functions are ideally suited to the application of digitally based techniques. The mobile phone plays a significant part in enabling progressive change for both subsets of agri-food decision-making.

An example of the first of these important areas of development is the implementation of a mobile-based customized advisory service for farmers (Ama Krushi) in Odisha, India by an organization named Precision Development (PxD).[5] PxD reached 5.7 million users in the fourth quarter of 2021. Ama Krushi is a free service provided by the Odisha Department for Agriculture and Farmer's Empowerment. It demonstrates how mobile phones enable access at scale, and farmer data enables customization, but PxD is clear about why its work has been successful: technology is effective when it is user-centric, dynamic, and iterative. One example of the payback is the work done on increasing the adoption of flood tolerant seeds. Two voice messages (delivered via the mobile phone platform) focused on highlighting the benefits of flood tolerant seeds that significantly increased adoption and knowledge among farmers with low land. There was:

- a 25% increase in adoption;
- a 7.5% increase in knowledge; and
- an estimated increase of ₹72 per hectare against a marginal cost of ₹4.1 ($1 = ₹81.52 on 19 November 2022)

Significantly, PxD envisages the transfer of intellectual property and the application of this innovative service to the local government in Odisha. The Ama Krushi Service was conceived as a build–operate–transfer model and PxD is working to hand over

---

[5]   Source: O. Mpinganjira. 2022. *Implementation of a Mobile-based Customised Advisory Service for Farmers of Odisha*, India [PowerPoint presentation]. Quoted in ADB. 2022. *Rural Development and Food Security Forum 2022*. Manila.

management and operations of the service to the government of Odisha and its designated partner. Thus, this digitalization example has delivered improvements in farming productivity and local capacity.

Whereas the work by PxD had a direct focus on producers in India, an example of digital technology that connects farmers and consumers at scale comes from the People's Republic of China (PRC). Pinduoduo is the largest agriculture-focused technology platform in the PRC. It offers a platform that connects farmers and distributors directly with consumers through an interactive shopping experience. Sixteen million farmers supply fruits and vegetables to Pinduoduo users and almost 900 million consumers use the platform. In 2020, the company turnover was $42 billion. Importantly, Pinduoduo has promoted and applied several innovations that improve farm productivity: advisory services and upskilling farmers, social commerce, and consumer-to-manufacturer connections. Farmers and consumers both benefit from the results of the Pinduoduo system: higher incomes for farmers and lower costs for consumers (Figure 3).

There are undoubted benefits to using digital hardware and software to transform the agri-food system. Digital technology is a true enabler, but care needs to be taken that the hardware and capabilities for its use are evenly distributed in the rural population and that there is no gender bias. Digital literacy is low for marginalized, food-insecure communities, and women and rural communities are often underrepresented in information systems. Digital infrastructure needs to reach those who can benefit the most, not just those who can most benefit from it commercially. This is a challenge

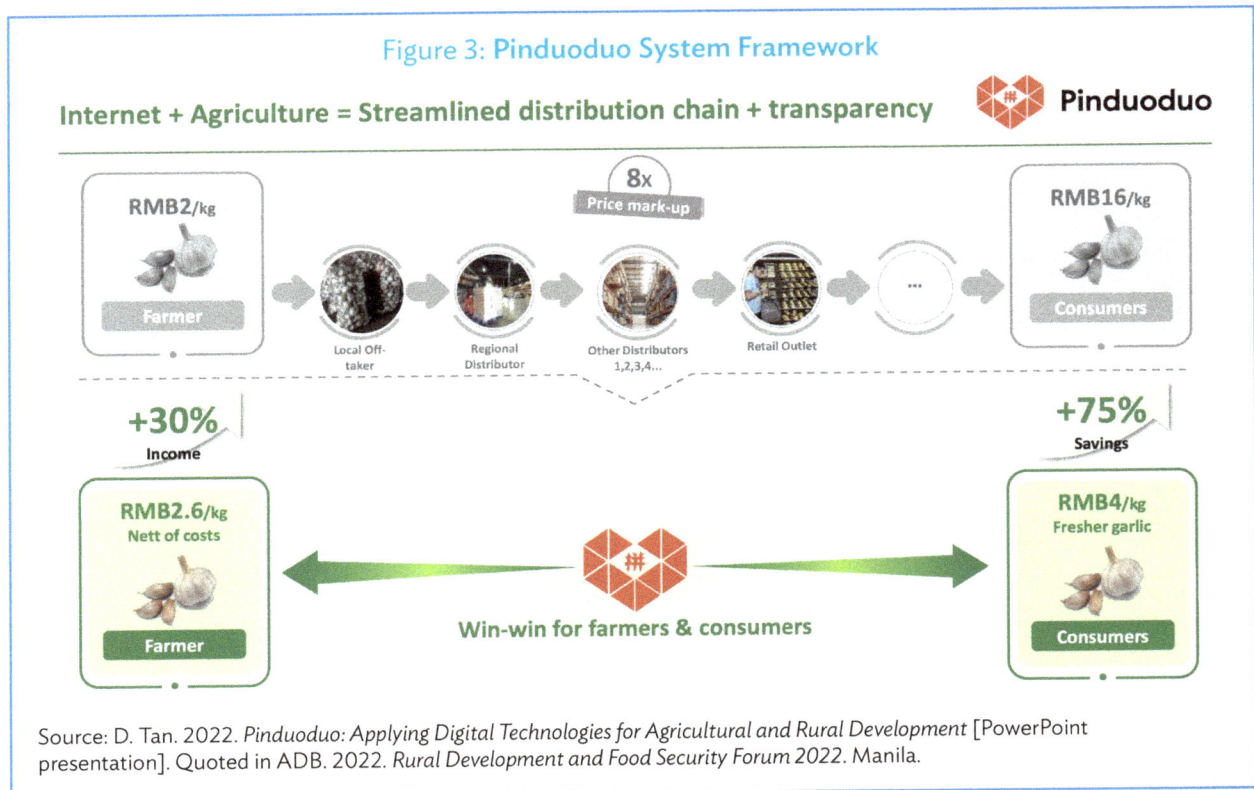

Figure 3: Pinduoduo System Framework

Internet + Agriculture = Streamlined distribution chain + transparency

Source: D. Tan. 2022. *Pinduoduo: Applying Digital Technologies for Agricultural and Rural Development* [PowerPoint presentation]. Quoted in ADB. 2022. *Rural Development and Food Security Forum 2022*. Manila.

for development banks and their partners, a recognition that the market may not deliver the tools of digitalization and subsequent benefits quickly enough without interventions. The productivity gains that digitalization offers may be enough to finance many of the investments in digital infrastructure and public financing may only be required for pump priming initiatives or for those communities where skills and resources are particularly constrained.

One final challenge for the digitalization approach should be acknowledged, and that is the need to update "traditional" systems of data collection, collation, and analysis used in developed economies and development banks and their partners. Decision-makers in government departments and development organizations also need to consider how to digitally transform their information systems to increase their productivity. Systems that are paper based—or that store data in "silos"—are an inefficient way to support evidence-based policy responses that can manage food system risks such as price hikes, pest infestations, floods, and droughts. Any bias in the accessibility of digital technology could lead to a bias in data collection and policy decisions—a danger that must be avoided. The potential gains for improved public information systems (especially for designing development finance and aid which are often remote from the market) are enormous. Real-time monitoring of data may be achieved, which may then facilitate more robust and timely analytics to plan and implement policy interventions.

III. Financing Sustainable Agriculture and Natural Capital

It is widely recognized that climate change materially erodes biodiversity, alters agricultural production, and threatens food security. On the other hand, expanding agricultural production accounts for up to 80% of biodiversity loss and is a major cause of climate change through GHG emissions. Further, shrinking biodiversity accelerates global warming and disrupts weather patterns, causing climate change with adverse effects on agriculture. Thus, agriculture production, biodiversity, and climate change have formed a vicious spiral that needs to be broken. Growing the base of natural capital to expand biodiversity could provide a sustainable solution to this problem.

Forests, agricultural land, the atmosphere, the oceans, and mineral resources are all examples of natural capital. These natural resources provide a range of ecosystem services—including food, water, energy, and shelter—which are vital for human life. Natural capital—which accounts for 20% to 55% of total national wealth—is a major driver of economic growth but it is hardly measured or considered in national budgets and investment decisions. An accounting framework for natural capital is needed.

Over millennia, humankind has exploited, polluted, or wasted the building blocks of natural capital, especially since the industrial revolution. These components of natural capital had inadequate regulation and protection—even though there might have been private property rights for them—or they were seen as public goods with low or zero protection from any authority. Their value was typically seen as zero since the market could set no prices for these "free goods"—a classic case of market failure. The correct response to this suboptimal situation is to introduce legally enforced standards to protect private property or—where there is no ownership—use laws and regulations that will enable the authorities to place

Box: The Natural Capital Project

The world's ecosystems can be seen as capital assets; if well-managed, their lands, waters, and biodiversity yield a flow of vital life-support services. Relative to other forms of capital, living natural capital is poorly understood and undergoing rapid degration. Often, the benefits nature generates are widely appreciated only upon their loss. The Natural Capital Project aims to change that paradigm.[a]

[a] Natural Capital Project. https://naturalcapitalproject.stanford.edu/who-we-are/
natural- capital-project.

values on the elements of natural capital and to protect them vigorously. In many cases, this protection will need cross border and multicountry cooperation.

Can this sequence of damaging actions (which produce climate change) be modified or explained using the traditional tools of accounting and macroeconomic analysis? The benefits—in terms of natural capital—can be estimated based on calculations of the present value of the additional flow of ecosystem services that meet the Sustainable Development Goals. Globally, the greatest gains—on average—come from investments in land remediation, followed by avoided deforestation, wetlands, materials efficiency, and air pollution reduction. But to turn this idea into a measurable and practical tool of development requires a theoretical model of how these environmental aspects can be involved in economic activity and investment decision-making.

The Natural Capital Project was launched in 2006 and is now supported by over 100 research institutions and many multinational agencies, including ADB. The Chinese Academy of Sciences, the University of Minnesota, the Stockholm Resilience Centre, The Nature Conservancy, and the World Wide Fund for Nature are partners in the initiative, which is centered at Stanford University in the United States. It has adopted the concept of gross ecosystem product (GEP) and Integrated Valuation of Ecosystem Services and Tradeoffs (InVEST). InVEST is a collection of open-source, free software models used to map and value the natural resources that sustain and enhance human life.

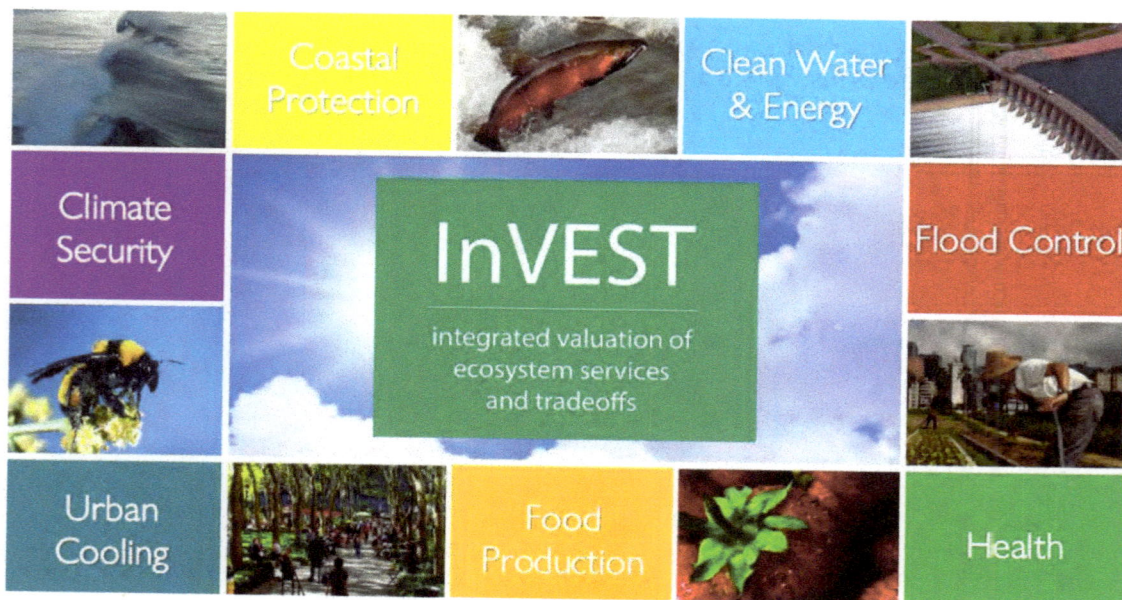

Figure 4: Integrated Valuation of Ecosystem Services and Tradeoffs Platform

InVEST = Integrated Valuation of Ecosystem Services and Tradeoffs.
Source: G. Daily. 2022. *Nature as an Engine of Prosperity* [PowerPoint presentation]. Quoted in ADB. 2022. *Rural Development and Food Security Forum 2022*. Manila.

Figure 5: Integrated Valuation of Ecosystem Services and Tradeoffs Framework

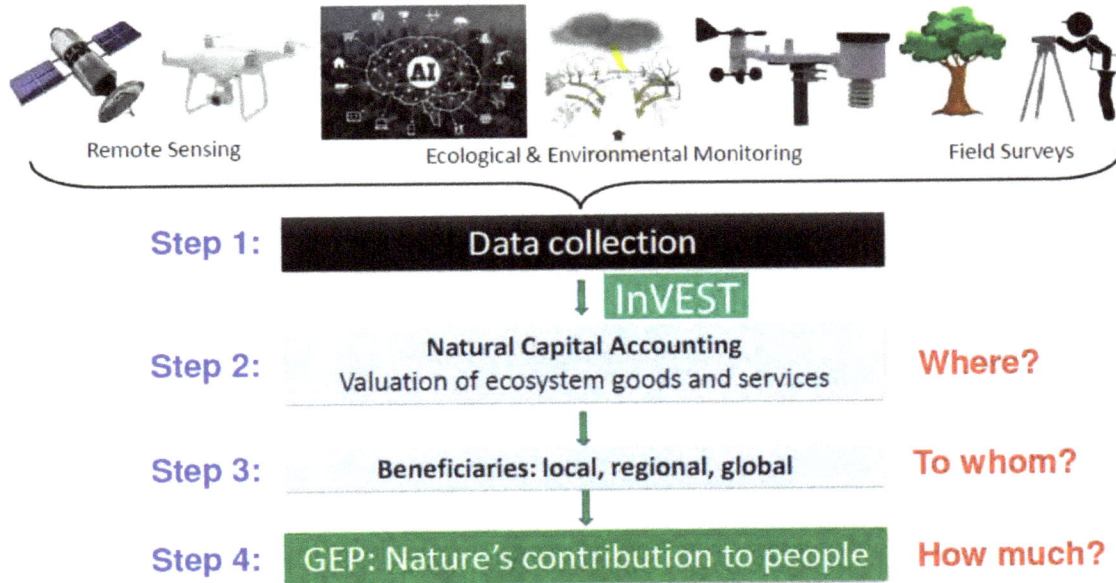

GEP = gross ecosystem product, InVEST = Integrated Valuation of Ecosystem Services and Tradeoffs.

Source: G. Daily. 2022. *Nature as an Engine of Prosperity* [PowerPoint presentation]. Quoted in ADB. 2022. *Rural Development and Food Security Forum 2022*. Manila.

Figure 6: Gross Ecosystem Product

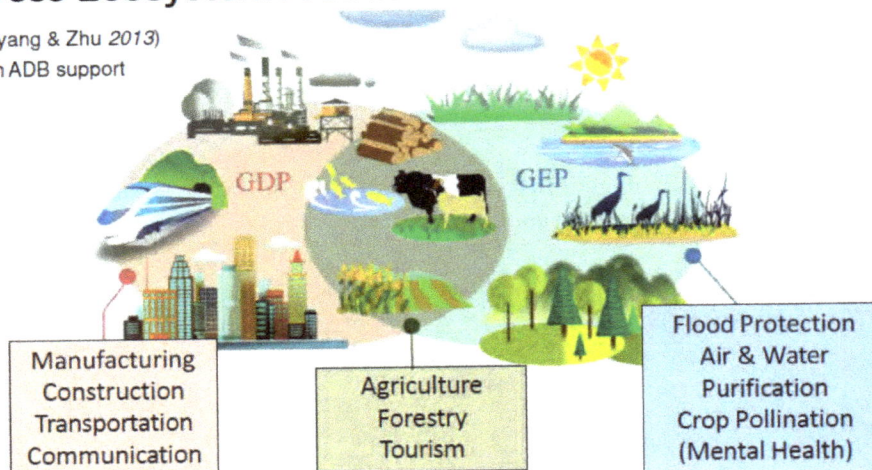

ADB = Asian Development Bank, GDP = gross domestic product; GEP = gross ecosystem product.

Source: G. Daily. 2022. *Nature as an Engine of Prosperity* [PowerPoint presentation]. Quoted in ADB. 2022. *Rural Development and Food Security Forum 2022*. Manila.

The multiservice, modular design of InVEST offers a useful tool for balancing economic and environmental objectives. InVEST allows decision-makers to evaluate the quantitative trade-offs related to alternative management options and to identify areas where natural capital investments can improve both human development and conservation. Maps are used as both information sources and outputs in InVEST models, making InVEST models spatially explicit. Results from InVEST are presented in both biophysical (e.g., tons of carbon sequestered) and economic terms (e.g., the net present value of that sequestered carbon).

This approach offers a bridge between accountants, financiers, environmentalists, civil society, and—importantly—the private sector in general. In addition to facilitating the implementation of sustainable changes, the acknowledgment of GEP enables the assessment of how changes in ecosystems can affect the flows of various costs and benefits to people and the environment. This is the future shape of development economics, regional and national technical assistance, and credit finance operations for major infrastructure.

The more progressive parts of the private sector exemplified by groups like Olam are trying to reimagine agri-food systems and actively participate in their transformation. Corporations need to assume the responsibility of regenerating soil, water, and ecosystems at large to enable smallholders and industrial farms to coexist. This is only possible if agriculture accompanied by this regeneration is financially rewarding. Revitalizing rural communities and enabling sustainable livelihoods while growing natural capital is a key aspect of meeting the challenge of feeding a growing population.

Adequate availability of financial capital is an important element of growing natural capital and breaking the vicious agriculture–climate–biodiversity spiral. Forests and agriculture hold over 30% of the climate crisis solution while receiving less than 3% of climate finance.[6] Innovative blended finance models that catalyze private capital towards sustainable agriculture that also protects land and water resources are required.

The AGRI3 Fund jointly promoted by Rabobank and the United Nations Environment Programme has begun a partnership for forest protection and sustainable agriculture to unlock at least $1 billion in finance. The fund intends to create business models that accelerate forest protection and reforestation, as well as the adoption of innovative agricultural techniques, all while improving the quality of life of local farmers and smallholders.

Recognizing the scarcity of finance for nature while leading the fight against food insecurity, ADB is designing a catalytic green blended finance facility called the Innovative Natural Capital Financing Facility (INCFF) to support projects with natural capital components. INCFF comprises three pillars: Natural Capital Lab, Natural Capital Fund, and Agribusiness Services Platform or Marketplace Platform (Figure 7).

INCFF would also create strategic partnerships—both technical and financial—for the scaling up of finance with greater impact.

---

[6]   Source: AGRI3 Fund. https://agri3.com/.

Figure 7: Innovative Natural Capital Financing Facility

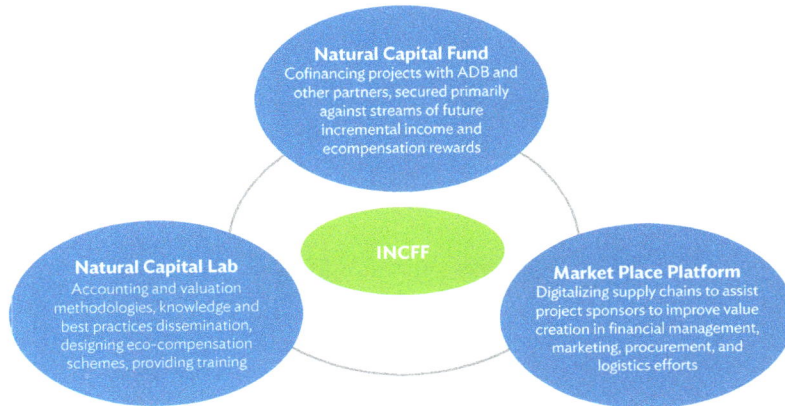

Natural Capital Fund
Cofinancing projects with ADB and other partners, secured primarily against streams of future incremental income and ecompensation rewards

INCFF

Natural Capital Lab
Accounting and valuation methodologies, knowledge and best practices dissemination, designing eco-compensation schemes, providing training

Market Place Platform
Digitalizing supply chains to assist project sponsors to improve value creation in financial management, marketing, procurement, and logistics efforts

ADB = Asian Development Bank; INCFF = Innovative Natural Capital Financing Facility.

Source: ADB. 2022.

Finance is a key engine of development. Sustainable transformation of agri-food systems requires the conservation and growth of biodiversity which, in turn, needs large amounts of financial capital. Private capital chases return. Thus, projects that generate returns and measurable conservation benefits need to be designed. Catalytic financing facilities such as the INCFF (with its Natural Capital Lab) can value ecosystem services and design payment for ecosystem services-type mechanisms, helping reduce overall investment risk and enabling private capital to fund natural capital-oriented projects.

Investments in nature are growing. However, the scale and pace of these investments must increase. Hopefully, innovative financing structures supported by natural capital valuation and concepts such as GEP can overcome investment barriers and enable much-needed capital to flow in.

IV. Nutrition Security

Global nutrition security is in an alarming state. According to the United Nations, the world is not expected to achieve targets for any of the major nutrition indicators by 2030. In 2020, there were 768 million undernourished people, and 418 million of those lived in Asia, according to the FAO State of Food Security and Nutrition Report for 2021.[7] The same report concludes "With less than a decade to 2030, the world is not on track to ending world hunger and malnutrition; and in the case of world hunger, we are moving in the wrong direction." While this undernutrition occurs, there is an obesity crisis in Asia. India, Indonesia, Pakistan, and the PRC, for example, occupy 4 of the top 10 countries with the highest populations of obesity in the world. Malnutrition can also be seen in Asia's adult population: 8.5% of women and 9.6% of adult males (aged 18 and older) have diabetes. At the same time, 6.0% of men and 8.7% of women suffer from obesity.

The nutritional challenge in many developing and emerging economies is a double challenge. The double burden of malnutrition (DBM): infant and child undernutrition occurring at the same time as overnutrition and obesity in the population. One of the highest percentages of malnourished children worldwide is found in South Asia, where micronutrient deficiencies such as a lack of iron, vitamin A, and zinc—as well as stunting, wasting, and being underweight—are particularly prevalent.[8] At the same time, Asia and the Pacific have the highest absolute number of overweight and obese individuals at 1 billion, with 2 out of every 5 adults being obese.[9]

These statistics are part of the evidence base that demonstrates why developing and emerging economies require strategies to improve food security, but also avoid having excess amounts of the "wrong food" in the system that encourages overeating for adults and the consequent ill health associated with obesity. Globally, there is now more obesity than there is undernutrition. In the developing world, obesity prevalence is catching up with undernutrition, while in the transitioning economies there is more obesity than there is undernutrition.[10]

---

[7]   Source: FAO, IFAD, UNICEF, WFP, and World Health Organization. 2021. *The State of Food Security and Nutrition in the World 2021. Transforming food systems for food security, improved nutrition, and affordable healthy diets for all.* Rome. https://doi.org/10.4060/cb4474en. Quoted in ADB. 2022. *Rural Development Food Security Forum 2022.* Manila.

[8]   Source: N. Wali, K. Agho, and A.M.N. Renzaho. 2019. Past drivers of and priorities for child undernutrition in South Asia: A mixed methods systematic review protocol. *Systematic Reviews.* 8 (189).

[9]   Source: ADB. 2018. *Fighting Obesity in Asia and the Pacific.* Manila.

[10] Source: United Nations Children's Fund (UNICEF), World Health Organization, International Bank for Reconstruction and Development/The World Bank. 2021. *Levels and trends in child malnutrition: key findings of the 2021 edition of the joint child malnutrition estimates.* New York: UNICEF. Quoted in ADB. 2022. *Rural Development Food Security Forum 2022.* Manila.

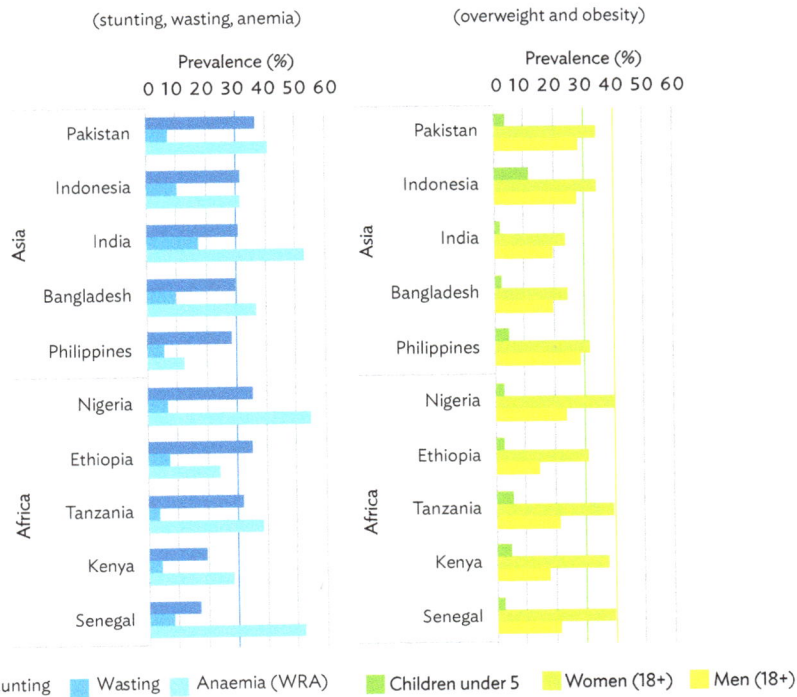

Figure 8: Distribution of Double Burden of Malnutrition across Countries

Source: M. Arabi. 2022. *Double Burden of Malnutrition, Double Duty Actions.* Quoted in ADB. 2022. *Rural Development and Food Security Forum 2022.* Manila.

The facts relating to DBM are startling and the scale and scope of the necessary solutions are imposing and urgent. There is much work to do. Most children with malnutrition live in Africa and Asia (Figure 8). In 2020, almost half of all children under five affected by overweight lived in Asia, and more than one-quarter lived in Africa. Since 2000, there has been a significant increase in the number of overweight children observed in Southeast Asia and northern Africa. An estimated 8.7 million children under five were overweight in selected countries, half of those children are in India and Indonesia.

Stunting remains the largest burden with 84.1 million children impacted and 30.8 million children wasted. Half of these stunted children and two-thirds of wasted children live in India. Adult women have a higher prevalence of overweight and obesity compared to men in all countries.

A. The Link between the Double Burden of Malnutrition and Noncommunicable Diseases

The double burden of malnutrition is the prelude to the rapidly growing burden of noncommunicable diseases (NCDs) that already account for 75% of deaths worldwide.

Early life undernutrition—as early as in utero—not only predisposes children to poor physical and cognitive development in life but also increased the risk of NCDs in adulthood. Since 1990, all-cause deaths due to NCDs have nearly doubled in Bangladesh, Ethiopia, Kenya, and Tanzania. A cross-over is seen where either overweight or obesity prevalence surpasses underweight adults. This is especially prominent for females. This precipitates further increases in the NCD burden among adults. The prevalence of overweight and obesity among teenagers is rising, and Asia and North Africa are seeing the fastest growth rates worldwide.

More research is needed to quantify the impact of early life undernutrition on later years. But it is already clear that—whatever the precise impacts—global conversations and research on the links between nutrition and NCDs should not be in separate channels. Joined-up thinking is needed to address these challenges.

B. The Socioeconomic Impact of the Double Burden of Malnutrition and Noncommunicable Diseases

If the health issues resulting from DBM and associated NCDs were not enough there are straightforward and easily justified economic reasons for addressing these issues. Poor health holds back the productivity and economic progress of a population, and this impact can be measured. Public investment in actions and policies to address DBM is likely to be encouraged if there are agreed, standard methodologies and metrics for estimating the economic impact of DBM. Cost-benefit analyses for these actions will be facilitated and the feasibility and payback from public investment will be better understood. The estimates of the impact of DBM on productivity and health-care expenditure that have been carried out indicate that developing countries have had—and will continue to bear—a growing economic burden from DBM. NCDs impose a wide and sweeping impact on society and the economy. Where NCD-related health care is not provided or cannot be readily accessed, one cost of NCDs on society is premature fatalities of productive citizens. Another cost—borne by individuals and families—is the crushing associated medical costs to deal with NCDs. On a national scale, health-care budgets will rise, and there is a loss of productivity and missed opportunities for the nation. But what do policies to tackle DBM and NCDs look like?

C. Double Duty Actions

Experts have suggested several interventions and programs that could potentially lower the risk or burden of both undernutrition (including wasting, stunting, and micronutrient deficiency or insufficiency) and overweight, obesity, or diet-related NCDs. (Figure 9). These so-called double duty actions illustrate the shared factors that underlie contrasting kinds of malnutrition. Double duty can be achieved at three levels:

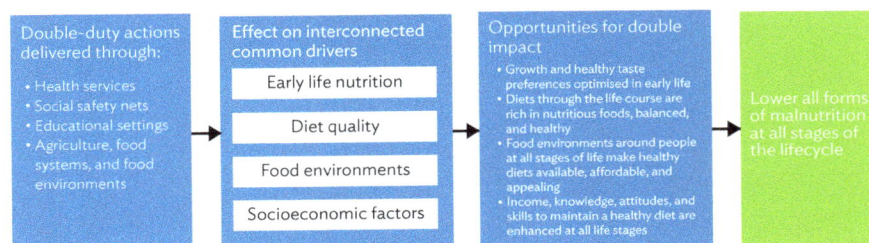

Figure 9: **Opportunities for Double Duty Actions**

Double-duty actions delivered through:	Effect on interconnected common drivers	Opportunities for double impact	Lower all forms of malnutrition at all stages of the lifecycle
• Health services • Social safety nets • Educational settings • Agriculture, food systems, and food environments	Early life nutrition Diet quality Food environments Socioeconomic factors	• Growth and healthy taste preferences optimised in early life • Diets through the life course are rich in nutritious foods, balanced, and healthy • Food environments around people at all stages of life make healthy diets available, affordable, and appealing • Income, knowledge, attitudes, and skills to maintain a healthy diet are enhanced at all life stages	

Source: C. Hawkes et al. 2019. Double duty actions: seizing programme and policy opportunities to address malnutrition in all its forms. *The Lancet.* 395 (10218). Quoted in ADB. 2022. *Rural Development and Food Security Forum 2022.* Manila.

- through doing no harm regarding existing actions on malnutrition;
- by retrofitting existing nutrition actions to address or improve new or other forms of malnutrition; and
- through the development of new, integrated actions aimed at the double burden of malnutrition.

Double duty interventions can be introduced by various agencies and organizations in the public and private sectors (Figure 9): health services, the social protection framework, education, and in agriculture and food production. Among nutrition interventions, the example of the benefits of prenatal micronutrient supplements is important. These have an impact beyond improving child survival and human capital. These supplements produce measurable reductions in the long-term risk of NCDs in the offspring generation. Also, the impact of calcium or multiple micronutrient supplements (MMS) at scale prenatally is equivalent to nearly half of the reduction in all-cause mortality attributable to high consumption of sugar-sweetened beverages in low and middle-income countries (32% of all deaths) estimated by the Global Burden of Disease. About 50% of deaths delayed were from ischemic heart disease and 30% were from stroke. A systems approach to these interventions is important, but before considering these aspects some examples from a public–private partnership that supplies these supplements and improved products are useful.

Examples of the impact of micronutrient supplements are available from the work of Royal Dutch State Mines (DSM). The World Health Organization listed MMS for pregnancy on its Model List of Essential Medicines in 2020. Royal DSM has partnered with various organizations to accelerate the delivery of MMS. Examples of these partnerships include

- Engagement with the Sight and Life Foundation, research, building scientific evidence, and executing trials, such as in Bangladesh;
- Enabling implementation research for MMS adoption in Nigeria along with UNICEF; and

- Enabling World Vision International to work on incorporating MMS to strengthen the health system of the Philippines, by improving the supply and demand of MMS for the last mile.

Royal DSM provides fortified rice kernels, rice flour, and cereals-based flour. These products all include multi-micronutrient supplementation MMS. In Rwanda, Royal DSM has invested in Africa Improved Foods (AIF). This is a public-private partnership involving Royal DSM, the Government of Rwanda, the International Finance Corporation, British International Investment Group, and Dutch Entrepreneurial Development Bank. AIF uses locally produced, highly nutritious meals to address malnutrition in a scalable and sustainable manner. Prime technology worth $65 million has already been invested in Rwanda and has been in use since December 2016.

Since 2016, AIF has reached more than 1.6 million consumers, sourced its inputs from over 130,000 smallholder farmers, and created over 300 direct jobs. It is estimated that AIF has contributed more than $1 billion in discounted net incremental benefits to the African economy.

Systems Approach

A system-based approach to nutrition programming is necessary for such a complex issue as DBM. It will facilitate strategies that can adapt to changing conditions and crosscutting pressures. The systems-led process promotes an understanding of how different elements are affecting holistic change rather than just understanding specific components in isolation.

Local systems need to leverage synergies, programs, and structures to promote and enhance practical sustainability. Positive and negative consequences that may occur need to be anticipated. Program planners need to be focused on what they can and cannot do. Partnerships with others to fill gaps should always be explored. Collaborative thinking should lead to coordinated actions. The behavioral challenge for policy makers is to consider how they move from silos to synergies.

The systems approach requires a shift away from fixed planning to more iterative and adaptive planning, and a focus on co-creation with local stakeholders. Context-specific solutions must be sought rather than generic ones, as paradigms and preconceived ideas often limit the understanding of local contexts. Clear pathways must be built for the integration of nutrition into systems, such as adapting education approaches to ensure the integration of nutrition. Assessments must be made on how to reach those not reached by current programs, such as how to reach the remaining 20% of adolescent girls (Figure 10).

The distributions of the multiple burdens of malnutrition and NCDs overlap but the interactions and interventions are also seen to be multilayered. An opportunity for a triple benefit exists if there is an investment in maternal and early childhood nutrition interventions i.e., actions that benefit one generation in addressing both undernutrition and overnutrition (double duty actions), and that carries through to

Figure 10: Conceptual Framework of System-Based Approach to Nutrition Programming

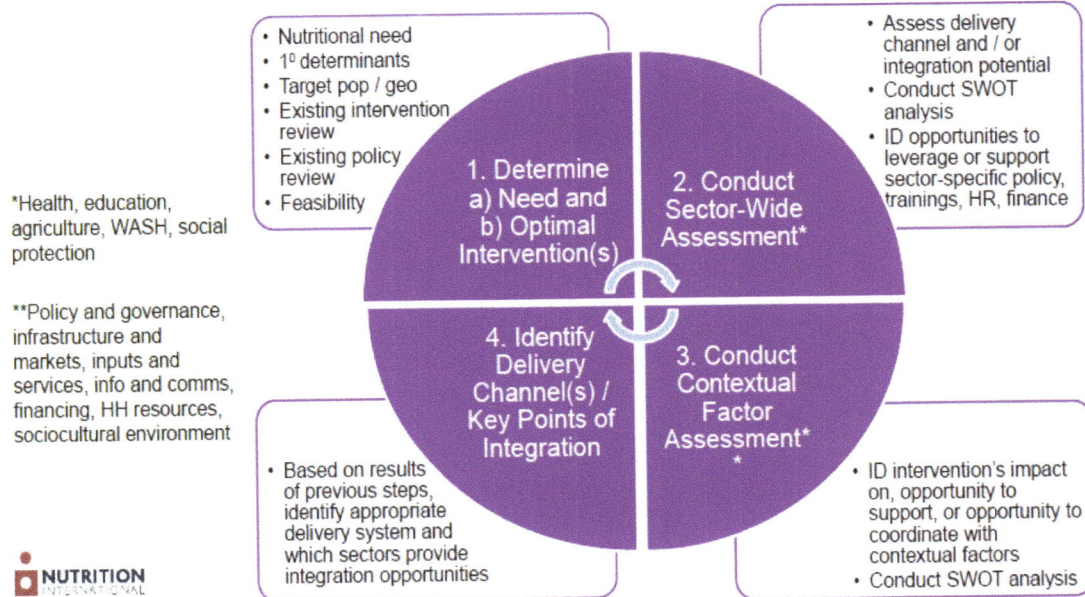

- Nutritional need
- 1⁰ determinants
- Target pop / geo
- Existing intervention review
- Existing policy review
- Feasibility

*Health, education, agriculture, WASH, social protection

**Policy and governance, infrastructure and markets, inputs and services, info and comms, financing, HH resources, sociocultural environment

1. Determine
   a) Need and
   b) Optimal
   Intervention(s)

2. Conduct Sector-Wide Assessment*

4. Identify Delivery Channel(s) / Key Points of Integration

3. Conduct Contextual Factor Assessment* *

- Assess delivery channel and / or integration potential
- Conduct SWOT analysis
- ID opportunities to leverage or support sector-specific policy, trainings, HR, finance

- ID intervention's impact on, opportunity to support, or opportunity to coordinate with contextual factors
- Conduct SWOT analysis

- Based on results of previous steps, identify appropriate delivery system and which sectors provide integration opportunities

NUTRITION INTERNATIONAL

ID = identify; HH = household; HR = human resources; SWOT = strengths, weaknesses, opportunities, threats; WASH = water, sanitation, and hygiene

Source: M. Arabi. 2022. *Double Burden of Malnutrition, Double Duty Actions* [PowerPoint presentation]. Quoted in ADB. 2022. *Rural Development and Food Security Forum 2022*. Manila.

the next generation (triple benefit). Data and robust, expert analyses are needed to identify the right mix of interventions, as well as leverage the right platforms for policy and decision-making. A systems approach and integration will produce synergistic benefits and savings at all levels of the pathway. What is clear is that nutrition is central to addressing the increasing burden of noncommunicable diseases. Nutrition needs to be prioritized as part of a larger focus on addressing all the NCD targets, and overall investments in health and human capital.

D. Collaborative Thinking and Actions: Linking Food Systems, Social Protection, and Health

The Pakistan Ehsaas Nashonuma Project (ENP) is an example of policy actions and a program of work that links poverty alleviation and nutrition for the poorest. As of 2022, the ENP serves 260,000 beneficiaries, and the Government of Pakistan and the World Food Programme are planning expansion to reach 1.5 million women and children in 131 districts by mid-2024. The program is being expanded with the support of the ADB.

Various studies have highlighted the socioeconomic cost of malnutrition, nutrient gaps, and non-affordability of nutritious diets e.g., malnutrition costs Pakistan $7.6 billion annually.

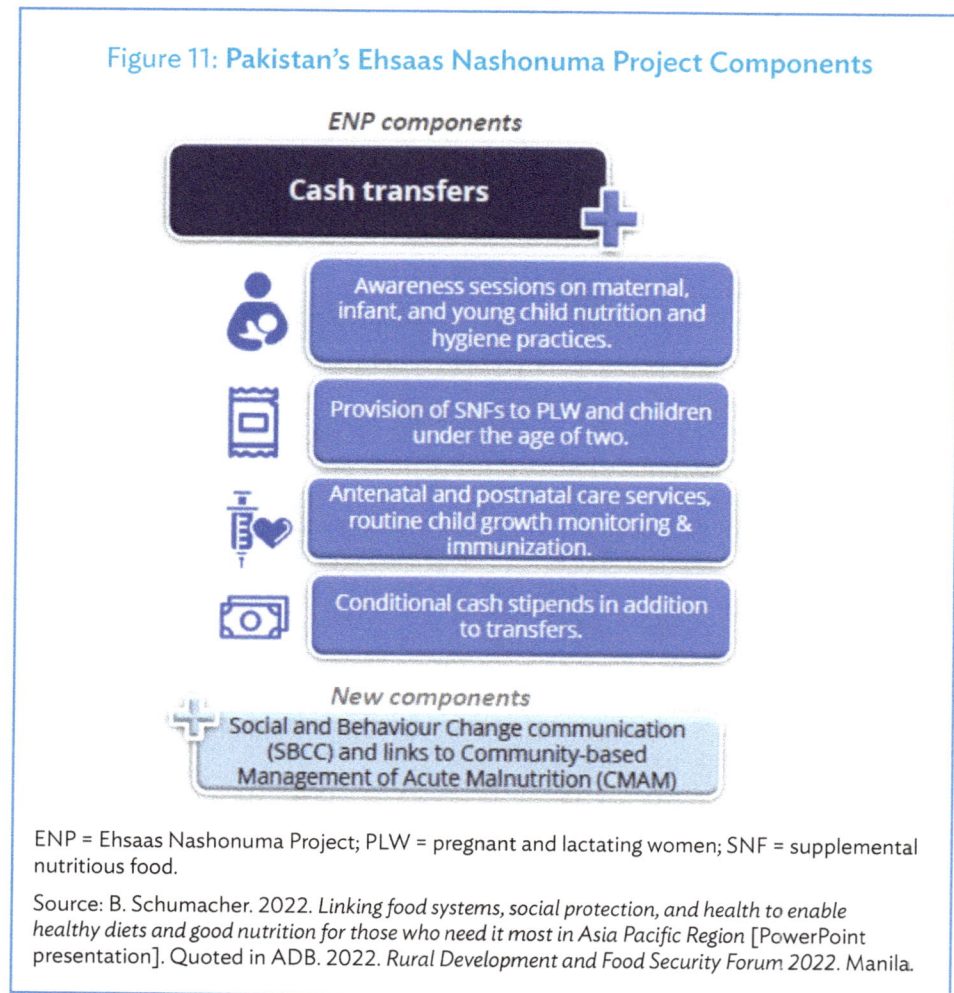

Figure 11 shows several components of ENP, demonstrating the more complex, systems approach that is recommended:

- Awareness sessions on maternal, infant, and young child nutrition and hygiene practices.
- Provision of supplemental nutritious foods to pregnant women and children under 2 years;
- Antenatal and postnatal care services, routine child growth monitoring and immunization; and
- Conditional cash stipends in addition to cash transfers.

It is clear from the ENP example that nutrition policy is about more than direct actions via a health department or agency, the challenge requires a holistic approach that will involve several arms of government.

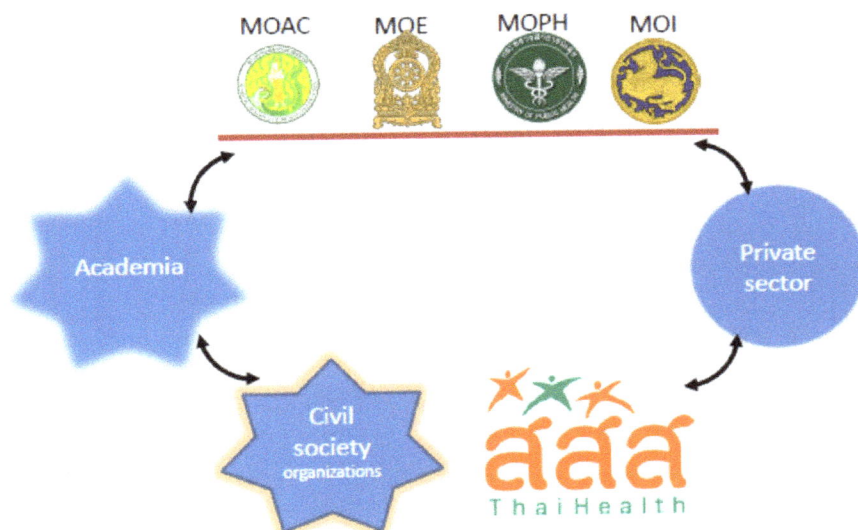

Figure 12: Thailand's 5-Year National Plan of Action for Nutrition

MOAC = Ministry of Agriculture and Cooperatives (Thailand), MOE = Ministry of Education (Thailand), MOI = Ministry of Interior (Thailand), MOPH = Ministry of Public Health (Thailand).

Source: L. Mo-suwan. 2022. *Intersectoral approach to nutrition security: Thailand.* Quoted in ADB. 2022. *Rural Development and Food Security Forum 2022.* Manila.

E. The Thailand Approach to Nutritional Challenges

In Thailand, several government agencies have been involved in improving nutrition in the general population. The Department of Health has introduced iodine and

iron–folic acid supplementation for pregnant and lactating women, iron supplementation for infants and children, and iron-folic acid supplementation for women of reproductive age. The Ministry of Education—in collaboration with the ministries of Interior and Agriculture and Cooperatives—has overseen a school lunch program and milk for all children in early childcare and kindergarten, and in primary schools a school lunch standard—the Thai School Lunch program—and the farm to school initiative. Regulatory interventions have included the Infant Food Marketing Control Act; sugar-sweetened beverage tax; front-of-pack labeling; nutrient profile model; restriction of marketing activities of high fat, salt, and sugar products; the FoodChoice app; and statutory food fortification.

The 5-year National Plan of Action for Nutrition in Thailand is explicit about the targets it is trying to achieve (Figure 12). In the 2019–2023 plan the target outcomes aim to reduce undernutrition, to have no increase in overnutrition, and to create a food and nutrition database.

The database will enable measures of the nutritional value of food in the country and the chance to create a food and nutrition surveillance system. Changing the behavior of consumers through better knowledge of nutrition is also an important target. There are some lessons for developing member learning in the Thailand example: outcomes are measurable, it is not too ambitious, and is monitored and reconsidered over relatively short time horizons (five-year plans).

F. Measuring Diet Quality

Measuring diet quality is not easy as it has many dimensions.

- **Digestibility of the food.** Influenced by food matrix (dietary fiber), processing, or preparation method.
- **Bioavailability of key nutrients.** Influenced by nutrient composition and presence of anti-nutrients.
- **Digestion capacity of the human gut.** Influenced by disease and/or environment, which influences nutrient absorption.

The properties and science of isotopes are offering researchers ways to measure these aspects of diet quality. Stable isotope techniques may be used to validate biochemical markers. The International Atomic Energy Agency is supporting coordinated research projects or technical cooperation projects that use nuclear techniques—including stable isotopes—to accurately assess different dimensions of diet quality.

Stable isotopes of iron and zinc have been used to assess the absorption of these micronutrients and to evaluate the efficacy of various programs like fortification and bio-fortification. For example, a study in India showed that pearl millet bio-fortified with iron and zinc covered the daily requirements for children. A stable isotope labeled vitamin A can be used to measure changes in vitamin A stored in the body because of interventions addressing vitamin A deficiency and to make sure the right amounts are being consumed. Another example from Indonesia uses isotopes to assess the impact of edible oil fortification with vitamin A. An isotope-based sucrose breath test helps with assessing sucrose digestion as an indicator of gut health. This technique is used in the Philippines to assess gut health considering high stunting rates. A stable isotope technique has been used in India, Thailand, and other regions to assess protein digestibility in legumes commonly consumed in those countries and will be applied in a new International Atomic Energy Agency-supported Asian regional project to assess how protein quality links to human health outcomes.

V. The Rural–Urban Divide

In almost every country, there is a gap between rural and urban dwellers, even within developed economies. This is often measured and reported in terms of incomes and average living standards comparing rural and urban areas but there is a long list of metrics that can be used to report on the quality of life and opportunities for advancement for rural dwellers and for those who live in urban areas. These comparative measures include poverty indexes, infant mortality, health, access to services, educational attainment, and numerous other variables. In less developed and emerging economies, the gaps between rural and urban dwellers tend to affect a much higher proportion of the population than in developed economies. And—unlike in developed economies—these deficiencies are likely to have a much larger impact on economic growth. They may also have an impact on climate change locally and nationally. Poor rural dwellers are more inclined to overstock their animals on pasture or cut down trees and forests, for example, to feed themselves. Hence, poverty in rural areas is characterized by local food insecurity and it contributes to national and global food insecurity. Climate change and protecting biodiversity are other reasons to reduce the rural–urban divide.

The factors that drive people to cities may have a sound economic basis and may be a necessary part of economic development at certain periods in the history of a country. But migration needs to be in balance with the opportunities for economic betterment and not just be a flight from conflict, famine, and social deprivation in the countryside. Even if the economic rationale is not immediately obvious, rural–urban migration decisions may be based on prospective access to schools and health care (which may have a long-term economic motive). These human capital endowments are likely to play a big part in decisions to migrate from rural areas in developing countries—or subregions—where the economic opportunities in rural areas are restricted. This suggests that reducing the rate of urbanization—which is often so high that it overwhelms the capacity of urban areas to absorb migrants—might be achieved by introducing and improving key services in rural areas. Education is one of those services.

> While all kids do not need to go to college, all children should be going to high school.
> -Scott Rozelle, Stanford University (Senior Fellow and Professor), Director, Rural Education Action Project (REAP)

A. Education and the Middle-Income Trap

The dangers of ignoring the development of human capital have been considered by researchers looking at how the transition from the middle income to the higher income category for a country depends upon human capital, principally, levels of education. The situation for Asia—and particularly the PRC and its rural population—has been examined by several academic researchers regarding data on educational standards, incomes, and employment. ADB has also provided a review of the experience of secondary education

Figure 13: **Share of Labor Force That Attained Upper Secondary Education, Middle-Income Countries**

Country	Share in 2015		
• Türkiye	36	**The Trapped**	We could add:
• Brazil	46		
• Argentina	42		1. India
• Mexico	34		2. Indonesia
• South Africa	32		3. Malaysia
			4. Pakistan
			5. Philippines
			6. Thailand
Average Middle Income	36		7. Viet Nam
			8. Others
OECD	74		
Middle income grads:	72		

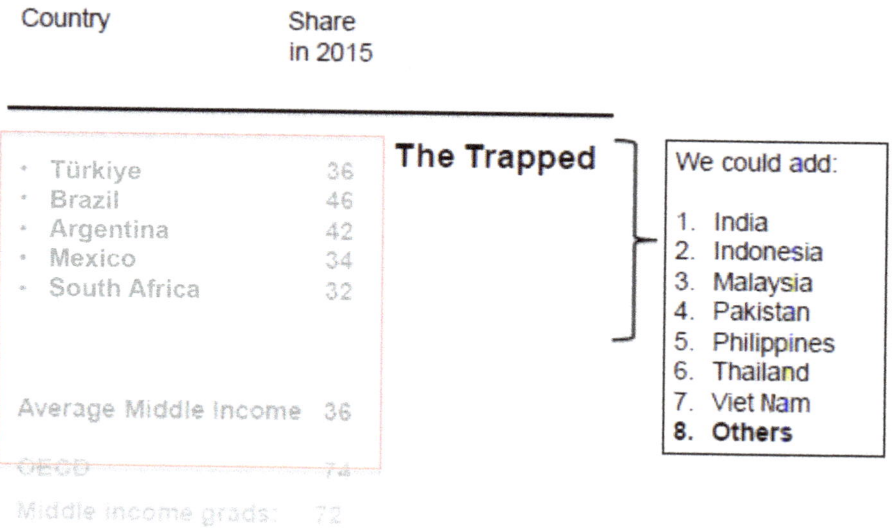

OECD = Organisation for Economic Co-operation and Development

Source: S. Rozelle. 2022. *Invisible Sectors: How Poor Levels of Human Capital Threaten the Rise of Asian Economies* [PowerPoint presentation]. Quoted in ADB. 2022. *Rural Development and Food Security Forum 2022*. Manila.

development in the Republic of Korea. The importance of universal secondary education, special policies and investment for rural areas, the investment in technical and vocational education and training high schools, and incentives for teachers who worked in rural areas is clear.

The transition from middle income to higher income economies depends on the levels of education in the entire labor force (as demonstrated in the Korean experience). It is critical at this (middle income) stage of development to get all children all the skills they (and the economy) will need in the future.

The nature of labor shifts from "low wage, low skill" to "high wage, high skill," as a country advances from medium income to higher income (Figure 13). Polarization happens when a sizable portion of the labor force is unable to participate. Along with supply-side issues of low productivity—such as an unfavorable investment climate, lack of skilled employees, increase of informal jobs, etc.,—demand-side issues such as high unemployment, low wages, high crime, and social unrest also arise. The argument is that the lack of education provision across a wide segment of society at the secondary school level will lead to middle-income countries not being able to attain high income status. Several countries appear to be so categorized: Brazil, India, Indonesia, Pakistan, the PRC, Türkiye, and other Asian countries are all "trapped." Investment in education and health is the only way out of the trap.

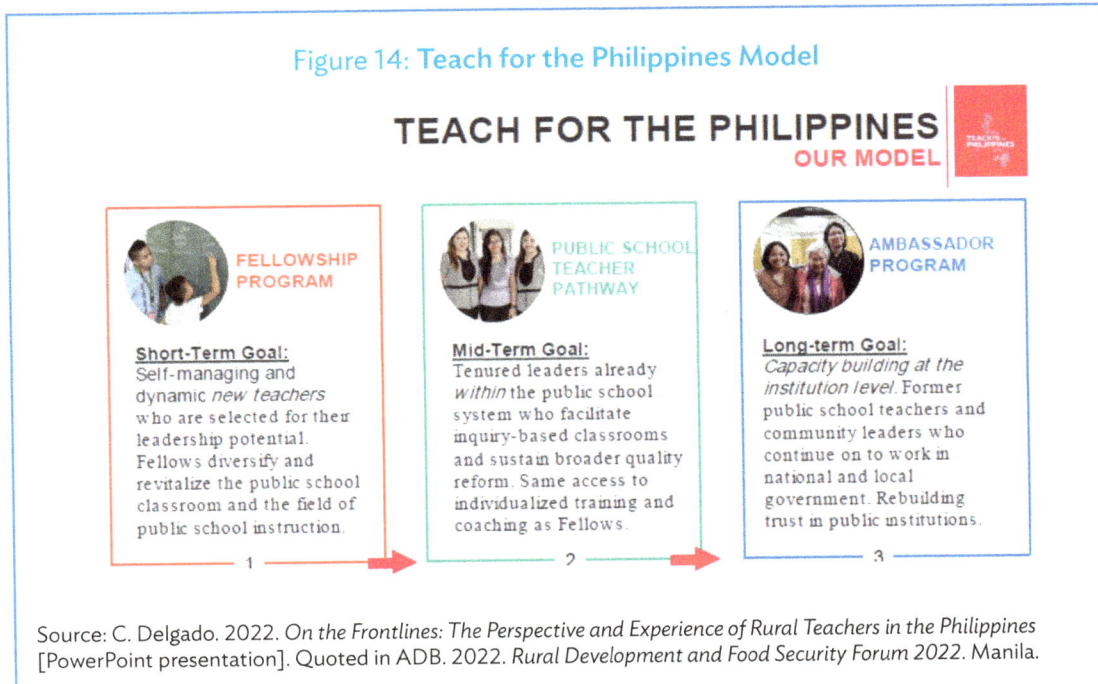

Figure 14: Teach for the Philippines Model

Source: C. Delgado. 2022. *On the Frontlines: The Perspective and Experience of Rural Teachers in the Philippines* [PowerPoint presentation]. Quoted in ADB. 2022. *Rural Development and Food Security Forum 2022.* Manila.

B. Education and the Rural–Urban Gap

The rural–urban gap—or divide—in developing countries is not a new phenomenon and the causes and drivers of these disparities have been researched and discussed for many years. Data from Bangladesh, India, and the PRC illustrate the divisions between urban and rural populations. Most schools in rural areas have teacher shortages, especially in English and Mathematics. Not all the teachers are fully trained, and rural schools have insufficient teachers, libraries, and laboratory facilities. Students from affluent urban households, however, can more afford coaching, private tuition, better guidance, and nutritious food than students from rural areas. Urban students have more advantages in most of the factors investigated in the PRC compared to their rural counterparts. Overall, the rural–urban gap in educational opportunities and attainment is far from satisfactory.

Investments in rural education programs can counter some of the negative factors that allow disparities between rural and urban populations to persist and examples from Bangladesh, the Republic of Korea, and the Philippines are informative. The Philippines also provides an innovative example of how food security and education about food production can be improved for low-income urban consumers.

The Teach for the Philippines program in the Philippines offers several insights into the challenges and solutions needed for development efforts in rural education (Figure 14). Providing education services in rural areas is not simply a matter of "copy and paste" from existing urban education programs. Teachers in rural areas tend to work harder to carry out their roles due to infrastructure challenges, societal beliefs, and a lack of resources for teachers and pupils. In many rural households, the value of education is deemed to be low. A rural teacher must help students not only appreciate the opportunities that education affords but often must also motivate families to support their child in finishing school. There is a scarcity of mobile and internet signals and gadgets that enable access. In Barangay Pacol, Kabankalan, Negros Occidental, the Teach for the Philippines program supplies teachers who train parents in storytelling and basic teaching methodologies. Due to the lack of mobile or internet access, during COVID-19 parents and learners benefited from home training from dedicated teachers to prevent learning loss.

Teachers in rural areas have a hard time accessing online learning resources either for their teaching or for professional development. Rural-based teachers may have to buy students their learning materials, like notebooks and pencils. Another example from the Philippines—in the pre-and post-pandemic periods—teachers in rural schools who had access to face-to-face training often have to spend their own money for transport fares to and from training venues, and in rural areas, these transport distances can be long. Teachers should not have to incur out-of-pocket expenses to provide pocket Wi-Fi devices to their students. In summary, all these factors make the provision of education in rural areas more difficult to carry out and more expensive. More money in the education budget, in the form of scholarships to students and/or subsidized transport and teaching materials, can overcome these issues at the micro level.

Rural schools tend to have smaller budgets. Providing more resources (money) is ideal if directly channeled to the schools rather than centrally from government agencies. Schools should be budget holders. Teachers and school leaders need to build professional skills outside of teaching. For example, training in resource and budget management as teachers and/or school heads often fundraise privately for community needs. Classrooms in rural schools are often not as well maintained as schools in urban areas. Parents in rural areas cannot always contribute to the repairs or maintenance of classrooms. Rural schools have higher costs, as their transport costs for the delivery of supplies and materials to a school will be higher than for the urban equivalent. Rural communities are more vulnerable to calamities resulting in greater losses of resources, time, and opportunities. These suggestions raise resource and governance questions: more money, training, and devolution of budgets to rural stakeholders are needed.

Less easily fixed are some of the infrastructure challenges that restrict education and other aspects of the rural economy. Physical infrastructure like public roads needs to be provided or improved in rural areas, and at least affordable and private transport services. Communication infrastructure is also needed. The internet enables so much in education and knowledge transfer and its access need to be a priority. Mobile and internet signals also need to be affordable, along with the gadgets (phones and laptops) that each family or student can use. At the macro level, rural education requires a whole-of-government approach. Alignment across agencies and departments can raise the quality of education in rural areas, but interdepartmental collaboration may be hard to achieve.

Figure 15: The Philippines' National Urban and Peri-Urban Agriculture Program

Source: W. Dar. 2022. *National Urban & Peri-Urban Agriculture Program: Responses & Prospects amidst Challenges in Food Security.* Quoted in ADB. 2022. *Rural Development and Food Security Forum 2022.* Manila.

C. Urban Farms

The Department of Agriculture in the Philippines has facilitated a program of innovation and education which has a focus on growing food in urban areas. The National Urban and Peri-urban Agriculture Program (NUPAP) seeks to address the challenges of urbanization, which have been exacerbated by the COVID-19 pandemic (Figure 15). The NUPAP program aims to improve food accessibility and availability and offer other sources of livelihood for urban dwellers using climate-resilient and sustainable urban agriculture technologies. The program will promote food safety for urban and peri-urban farming and engage and revitalize urban communities.

There are multiple advantages of these urban farming systems:

- Improves food access among low-income communities in urban areas.
- Improves diversification of crops and vegetables, products are consumed by the producers, or sold in retail markets.
- Provides access to local, fresh, and nutritious foods and products is increased, encouraging farmers' markets and the involvement of producers in marketing.
- Reconnects with the community through food, skills, jobs, and economic development.

One private sector company that is involved in NUPAP is Agrosheriff. Agrosheriff offers open field irrigation systems, custom-built greenhouses, optimized hydroponics and fertigators (an automatic fertilizer system connected to an irrigation system), and watering system kits and equipment. One of its clients—Growtech Farms—

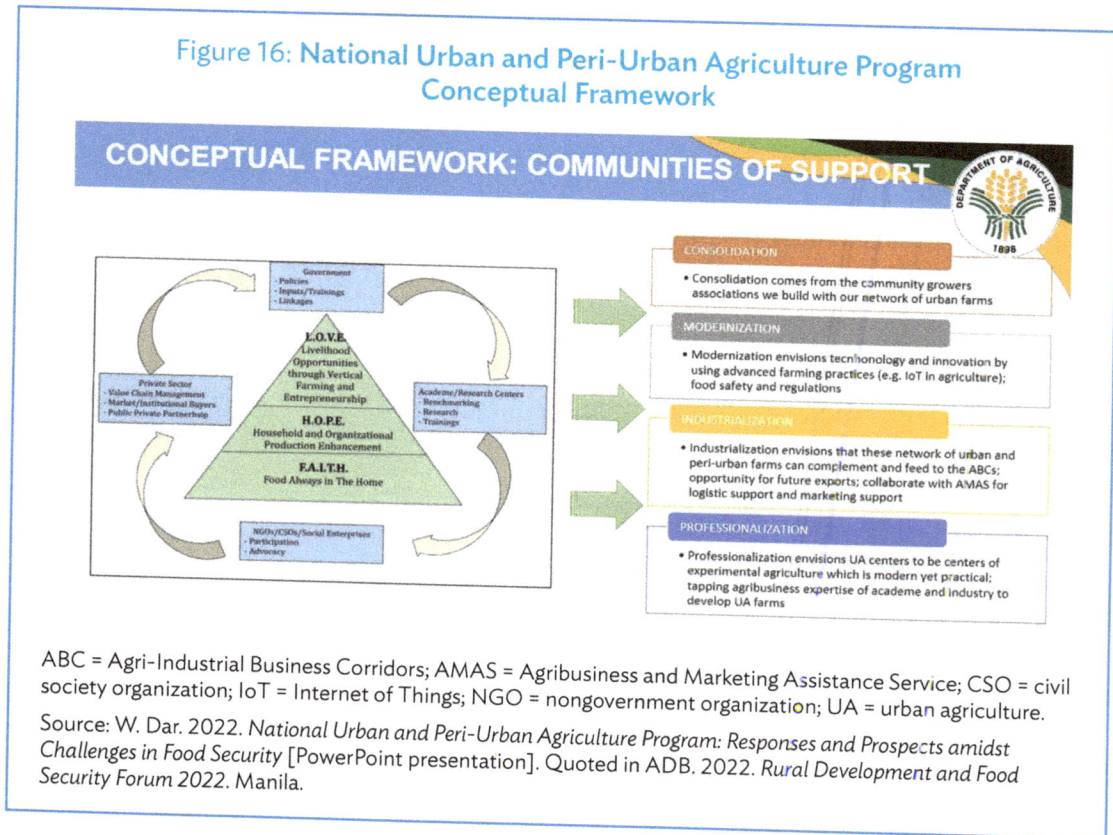

Figure 16: **National Urban and Peri-Urban Agriculture Program Conceptual Framework**

ABC = Agri-Industrial Business Corridors; AMAS = Agribusiness and Marketing Assistance Service; CSO = civil society organization; IoT = Internet of Things; NGO = nongovernment organization; UA = urban agriculture.

Source: W. Dar. 2022. *National Urban and Peri-Urban Agriculture Program: Responses and Prospects amidst Challenges in Food Security* [PowerPoint presentation]. Quoted in ADB. 2022. *Rural Development and Food Security Forum 2022.* Manila.

has initiated a pilot urban farm in Novaliches, Quezon City in the Philippines. The facility uses Israeli agricultural equipment and technologies—such as hydroponics and irrigation fertilizer systems—to cultivate strawberries, lettuce, and other crops. Agrosheriff aims to improve food security in the Philippines, improve the quality of fruits and vegetables, reduce imports, increase high-tech jobs, and increase the interest of young people to work in agriculture. One project is the creation of an urban farm school, a joint effort between the Department of Agriculture and Agrosheriff.

Indoor vertical farms offer the production of high-quality fresh food that does not require pesticides and meets high safety standards. There is no fertilizer runoff into the ecosystem when the hydroponic system is used, and labor inputs are lower than in conventional farms. Importantly, these vertical farms can be in urban areas where the need for fresh food and community involvement in food production is high.

NUPAP aims to build climate resilient and sustainable urban agriculture technologies and promote food safety for urban and peri-urban farming (Figure 16). This will not just deliver food but offers the chance to rehabilitate communities through urban farming since jobs and skills will be offered as part of a drive to bring food production into urban areas. Access to nutritious, fresh food will be improved and logistics costs will be reduced as food will be locally produced. NUPAP is a collaboration with local government units, national government agencies, international organizations, private sectors, social enterprises, marginal groups, and other interested stakeholders.

VI. Conclusions – The Way Forward

The challenges facing developing and emerging economies are growing, and they are worldwide. Since 2000, there have been signs that a global consensus that seeks unity and cooperation across borders to deal with these challenges has emerged and can be sustained. Collaboration and working together are often difficult for children to conceive and nurture but these cooperative forms of action are showing signs of life and maturity. The meetings and announcements associated with major intergovernmental organizations such as the 2030 Agenda for Sustainable Development, the Paris Agreement, and the United Nations Climate Change Conference have gathered pace. And all are spawning guidance and prescription on the realignment of policy proposals, targets, and reforms and putting evidence-based interventions into practice.

A global effort is underway to redirect the economic and environmental activities of humanity. The development community (donor countries, agencies, banks, and partners) is playing its part in this redirection, which has been demonstrated by the contributions at the 2022 forum. The community has offered targets, resources, and practical ideas to improve economic and human development, enhance food security, and adapt to and mitigate climate change. But there is much to do, not least because some new challenges have arisen, and others have become more urgent. For example, extreme weather events seem to be arriving more frequently and becoming more extreme. The impact of poor nutrition on human health and productivity has been recognized for decades but the double burden of malnutrition is now seen to be very relevant to climate change policies as well as to healthy lives. This is an issue that particularly affects the Asia and Pacific region.

COVID-19 has reminded people and policymakers that pandemics are a risk factor that can throw the plans of governments and development practitioners off course. While conflicts, war, and political upheaval may have been confined to some subregions and populations, they have been shown to have wider global impacts. They are another risk factor that is now increasingly important in the development environment and they further emphasize the need for resilience and sustainability when faced with external shocks.

The 2022 forum has shown that challenges are growing and are increasingly complex in nature and impact. A common reaction to these challenges has been to recommend and design policies that improve human capital and that work across several dimensions: health, food policy, education, social protection, the environment, gender equality, etc. This is an observation that helps form an initial conclusion about the way forward for climate change policies. The future for interventions that transform agriculture, improve food security, reverse climate change, and improve nutrition and health will be a holistic one. Multidisciplinary efforts and collaborative actions by governments and development partners are essential. There should be no more siloes and much more coordinated thinking. This holistic approach brings further complexity as it implies that development partners, government departments, donor countries, and developing and emerging economies need interventions that will require design, communication, and management skills that explicitly aim to be "joined up" vertically and horizontally. The skills needed to achieve this will not be acquired by accident. Increasing the capacity to invest in and apply these skills and attributes is the conclusion that follows from this discussion. Being "joined up" needs explicit recognition and support in development programs through the appointment of "holistic champions." Successful development outcomes in individual programs have often resulted from the presence of such a champion(s), sometimes by accident rather than design.

Market signals and private sector operators can be powerful forces in identifying issues and deploying resources to meet demands and help transform supply chains. Private firms provided examples of producers improving productivity and prospering from knowledge and innovation at the forum. The market is a source of resources and guidance and offers inclusivity, subject to market participants having finance. The purchasing power of smallholders and consumers may be small at an individual level, but even relatively small amounts of producer-sourced finance or consumer-driven demand can leverage resources and change at a local level: especially if the market is innovative about how producers and consumers can interact and act collectively.

The digitalization of the agri-food chain is a prime example of how technology has helped agripreneurs and consumers transform some value chains in Asia, as demonstrated at the forum. Small and medium-sized enterprises and smaller companies have been able to play a key role in transformation efforts, not just multinationals and larger domestic companies. Smaller firms and agripreneurs in the private sector will often be the early adopters and pioneers of innovative systems for agricultural production and distribution. However, they often suffer from barriers to entry, unequal market power, and distorting regulations. Enabling and leveraging market forces and private sector operators—in all forms and sizes—needs to be seen as a desirable strategy to effect change and transformation. Facilitating actions by the development community may take the form of providing infrastructure (smartphones, Wi-Fi, broadband, etc.) and programs that deliver knowledge, training, and education. In parallel, identifying and supporting agripreneurs and progressive private firms are aspects of developing human capacity. Producers and economic actors are on the front line of development and can be enormously powerful in spreading ideas and innovation, especially if stakeholders and development partners offer support to achieve scale.

Resilience and risk are aspects of development that may have been underplayed. There have always been vulnerability to natural hazards—earthquakes, floods, droughts, crop failures, and livestock diseases, but the world now seems to be operating in a different, higher range of values for risk. Climate change is the main cause of extreme weather events, and disease outbreaks are encouraged by the exploitation of natural resources, population growth, and urbanization. Further, social upheaval and conflict increase the risk profile of all investments and commercial activity. This new, riskier economic environment suggests that planning for resilience will be a more important part of delivering food security in the future. Development partners and stakeholders may need substantive ways of dealing with risk and uncertainty in the future, or increased budgets for contingencies. The increasing and climate-induced volatility cannot be dealt with by smallholder farmers.

One aspect of risk that deserves separate mention is social protection. COVID-19 has brought this issue into focus as finance ministers around the world realized that the pandemic and its associated lockdown policies had the potential to bring economic and social catastrophe even in industrialized countries. The experience of COVID-19 is a reminder that an economic system will struggle if there are not enough people to maintain and manage it. Another positive aspect of the pandemic is the growing understanding that measures to improve social protection in certain circumstances has grown. The conclusion from this is the acceptance that budgets and programs may need social protection measures as part of their design.

Green finance and the relatively new subject of natural capital are other areas of thinking and work that will profoundly change calculations about development activity and program design. The widespread adoption of the concepts of gross ecosystem product and InVEST will be transforming. These ideas will be a necessary condition for future decision-making by all stakeholders, in and outside of the development community. Despite the huge challenges that the developing world is facing, there were many significant examples of progress and best practice in the 2022 forum.

Those examples—from the private and public sectors and partnerships—are for all stakeholders: leaders, agripreneurs, governments and agencies, investors, smallholders, and multinationals. These frontline success stories came from all parts of the Asia and Pacific region and the case studies offer lessons from the field that are valuable pointers for how to transform agri-food systems and meet the challenges of climate change, nutrition, and the rural–urban divide.

In addition to strengthening its ongoing efforts, ADB has enhanced its cooperation with other partners to implement recommendations and suggestions made in the forum. ADB and the Ministry of Agriculture, Forestry, and Fisheries of Japan signed a memorandum of cooperation in September 2022 to work together to promote sustainable, resilient, and inclusive agri-food systems in the Asia and Pacific region by (i) extending innovative agricultural production and marketing technologies adapted to climatic and agronomic conditions in developing member countries (DMCs); (ii) promoting sustainable and climate-resilient natural resource management practices; and (iii) enabling institutions to support the sustainable and resilient transformation of agri-food systems in DMCs. ADB is also expanding the areas of cooperation with the International Atomic Energy Agency to promote the supply of safe and nutritious food. To scale up nutrition-smart investment, ADB has started working with the International Food Policy Research Institute and HarvestPlus.

ADB is also working with International Rice Research Institute to increase climate financing and provide science-based solutions for building sustainable and resilient food systems. It is also working to build a partnership with the World Food Programme to promote food security. ADB is working to develop an Innovative Natural Capital Financing Facility to enhance investments in sustainable food value chains, climate resilient agriculture, and natural resource management. This facility will integrate nature-positive solutions to project design by (i) applying natural capital accounting methodologies to value natural assets and computing GEP projects, (ii) preparing projects from a natural capital perspective by designing eco-compensation schemes to improve project viability and mitigate risks, (iii) determining regulatory frameworks in the region at the DMC level, and (iv) building partnerships and enhancing institutional capacity by disseminating knowledge among institutions. ADB is also working to strengthen cooperation to promote agricultural and rural development with its bilateral partners like the Republic of Korea and Japan.